Geobotany Studies

Basics, Methods and Case Studies

Series editor

Franco Pedrotti, University of Camerino, Camerino, Italy

The series includes outstanding monographs and collections of papers on a range of topics in the following fields: Phytogeography, Phytosociology, Plant Community Ecology, Biocoenology, Vegetation Science, Eco-informatics, Landscape Ecology, Vegetation Mapping, Plant Conservation Biology, and Plant Diversity. Contributions should reflect the latest theoretical and methodological developments or present new applications on large spatial or temporal scales that will reinforce our understanding of ecological processes acting at the phytocoenosis and vegetation landscape level. Case studies based on large data sets are also considered, provided they support habitat classification refinement, plant diversity conservation or vegetation change prediction. *Geobotany Studies: Basics, Methods and Case Studies* is the successor to *Braun-Blanquetia*, a journal published by the University of Camerino from 1984 to 2011 in cooperation with the *Station Internationale de Phytosociologie* (Bailleul, France) and the *Dipartimento di Botanica ed Ecologia* (Università di Camerino, Italy) and under the aegis of the *Société Amicale Francophone de Phytosociologie*, the *Société Française de Phytosociologie*, the *Rheinold-Tüxen-Gesellschaft* and the *Eastern Alpine and Dinaric Society for Vegetation Ecology*. This series promotes the expansion, evolution, and application of the invaluable scientific legacy of the Braun-Blanquet school.

More information about this series at http://www.springer.com/series/10526

Alejandro Velázquez · Consuelo Medina García
Elvira Durán Medina · Alfredo Amador
Luis Fernando Gopar Merino

Standardized Hierarchical Vegetation Classification

Mexican and Global Patterns

Springer

Alejandro Velázquez
Centro de Investigaciones en Geografía Ambiental
Universidad Nacional Autónoma de México
Morelia, Michoacán
Mexico

Alfredo Amador
Facultad de Biología
Universidad Michoacana de San Nicolás de Hidalgo
Morelia, Michoacán
Mexico

Consuelo Medina García
Centro de Investigaciones en Geografía Ambiental
Universidad Nacional Autónoma de México
Morelia, Michoacán
Mexico

Luis Fernando Gopar Merino
Centro de Investigaciones en Geografía Ambiental
Universidad Nacional Autónoma de México
Morelia, Michoacán
Mexico

Elvira Durán Medina
Centro Interdisciplinario de Investigación para el Desarrollo Integral Regional
Instituto Politécnico Nacional
Oaxaca, Oaxaca
Mexico

ISSN 2198-2562 ISSN 2198-2570 (electronic)
Geobotany Studies
ISBN 978-3-319-41221-4 ISBN 978-3-319-41222-1 (eBook)
DOI 10.1007/978-3-319-41222-1

Library of Congress Control Number: 2016943442

Printed on acid-free paper

This Springer imprint is published by Springer Nature
The registered company is Springer International Publishing AG Switzerland

Apostolado en las Alturas

Ernesto Velázquez Hurtado†

A quien se esfuerce en subir a la montaña
A quien conoce el aire plácido de las Alturas
A quien sudando entre el hielo, la nieve y las alburas
A quien se encuentre esclavo con la hazaña
A quien practique el sermón en la montaña

A quien entienda lo dicho y lo comulgue
A quien a esfuerzo titánico se humille
A quien a esas alturas se arrodille
A quien con valor la verdad divulgue.

To Jerzy Rzedowski and Graciela Calderón who devoted more than half a century to inspiring generations of emerging naturalists to increase the knowledge of botany worldwide. Jerzy and Graciela planted the seed in the authors of this book by making us understand that botanical studies were unthinkable without remote sensing tools. Their vision was innovative and stimulating and we simply express our appreciation and admiration by means of this humble contribution.

Foreword

During this past decade, there have existed two realities in the study of vegetation. On the one hand, there has been an enormous quantity of information amassed, and, on the other, applying any of the existing classification systems to specific conditions in Mexico can be difficult. This has led to reflection, upon analysis, revision, readjustment, and specification, on a newly proposed vegetation classification system, the adoption or rejection of which will serve as an indication of its success.

The first chapter of this work consists of an historic review of the prevalent approaches to the study of vegetation both in Mexico and in other regions of the world. The second chapter analyses succinctly the existing schools of thought which have led to the development of vegetation classification systems based on physiognomic, structural, and floristic approaches.

The nub of the work is found in the third chapter, in which the proposed —Standardized hierarchical Mexican vegetation classification system‖ (SECLAVEMEX—Sistema jerárquico estandarizado para la clasificación de la vegetación de México) is presented. Organizational levels are laid out along with the criteria defining them and the nomenclatural basis for the denomination of each type of vegetation. Also included is a series of tables explaining and defining precisely the meaning of each concept, criterion, character, and element used with a view to assisting the reader in successfully identifying the type of vegetation in a determined area.

The fourth chapter highlights SECLAVEMEX's inclusive character as evidenced through its compatibility with systems currently prevalent in the Western world and in the tropics. The fifth chapter deals with a core current issue regarding the mixed approach to study land cover/land use/vegetation patterns. This chapter clarifies the (dis)similarities among concepts, tools and advantages of studying vegetation from an inclusive approach where these three concepts and used coherently towards a common goal.

There can be little doubt that the danger to plant life has moved on from being an alarming rate of extinction in terms of individual species to being one of an accelerated rate of habitat loss. In both cases, a good stocktaking is increasingly urgent, while, equally as crucial, is gathering information on vegetation from the perspective of having an adequate knowledge of natural resources. In Europe, for example, thanks to the adoption of the phytosociological approach more than a century ago as a basis for the study of vegetation, today, there are strategies for the management and conservation of habitats. In Canada and in the USA, the task was undertaken only a few decades ago, and, today, there are already promising results.

An additional motive behind this work is the urgent need to define universal criteria with the aim of creating a common tool in the struggle to preserve natural heritage, not only for the sake of knowledge, but also for future generations.

We, the authors, are convinced that this work is not by any means complete and that it will be the task of those who decide to try to use SECLAVEMEX to refine and adjust it to particular situations. This current work is but an effort to contribute to a rigorous yet user-friendly technical–scientific classification and inventory framework for all types of vegetation in Mexico.

Acknowledgments

We thank the Academic Unit of the Institute of Geography of the National Autonomous University of Mexico (now the Centre of Research in Environmental Geography) for its boundless support towards the accomplishment of this project. José Luis Palacio at the start, and Gerardo Bocco at the conclusion, maintained this institutional support. UNAM-Canada throughout Federico Fernández and Ramón Peralta provided support in the last phase of this book. Rigorous reviews of an earlier version of the book were made by Jerzy Rzedowski, Exequiel Ezcurra and Guillermo Ibarra, and, last but not least, Franco Padrotti, who all devoted outstanding time and energy to help us pursue this endeavor. Discussions took place over a decade with Orlando Rangel, Joaquin Giménez, Arturo Victoria and Sergio Zamudio, all globally-known botanists, whose critiques helped towards building a fruitful outcome. The support of colleagues such as Alejandra Larrazábal, Faustino López, Rocío Aguirre and Azucena Pérez, are all much appreciated. James Fouzie was responsible of the translation from Spanish into English.

A number of significant people have contributed their special gifts to help accomplish this book. The first author (Alejandro) could have not found greater motivation than from his wife Cecy Salgado and his son Erick Velázquez who in the past decade have supplied endless and warm support. The second author (Consuelo) found her motivation from her nephews. Amador acknowledges his wife Laura Torres and his students for pushing him forward in difficult times. To all of them, our most sincere appreciation and love.

Contents

List of Figures

List of Tables

List of Appendices

Prologue

In this volume, Alejandro Velázquez, Consuelo Medina, Elvira Duran, and Alfredo Amador and Luis Fernando Gopar (who work in three research centres of the Universities of Morelia and Oaxaca in Mexico) describe a new system of classification of the vegetation of Mexico, SECLAVEMEX (*Standardized Hierarchical Mexican Vegetation Classification System*).

The classification of vegetation cover has always been and remains one of the most important themes in the study of vegetation. The goal of classification was to establish groups, or classes, that cover relatively homogeneous phenomena in wide areas. This goal can be achieved on the basis of various criteria, such as physiognomy and structure, species and ecology, biogeography, dynamism, and genetics, as seen in the proposals of the succession of authors from 1800 to the present, among whom it is well to mention Brockmann-Jerosch, Rübel, Diels, Huguet del Villar, Sukaciov, Sochava, Dansereau, Küchler, Gaussen, Fosberg, Beard, Ellenberg, and Mueller-Dombois.

Chronologically, the first proposals referred to the physiognomy and structure of the vegetation; over the years as our geobotanical knowledge about the earth advanced, they were substituted by more complex and detailed proposals that also took into consideration ecological and floristic aspects.

Vegetation classification can be done deductively, moving from higher hierarchical units to lower ones, or inductively, when one starts from lower units and progresses towards the higher ones. The typological system of classification of plant formations of Brockmann-Jerosch and Rübel (1912) is deductive, while the floristic ecological (or phytosociological or syntaxonomical) system of Braun-Blanquet (1928) is a typical case of an inductive system.

The map of the natural vegetation of Europe on a scale of 1:2,500,000 (Bohn et al. 2000) was based on 19 formations, 14 of which were considered zonal and extrazonal, depending primarily on climate. They represented the main marocimatic zones and belts, while the other five formations represented azonal vegetation, depending on specific soil and hydrological conditions. Thus, the formations were classified into subgroups according to their species composition, finer climatic gradations, and variations in habitat conditions, and these were again subdivided according to nutrient levels, altitude, water budget, and geographic location. In practice, every subgroup is comprised of one or more similar plant associations. The classification used for the map of the natural vegetation of Europe was thus obtained starting from lower categories (associations and groups of associations) and grouping them into increasingly broader ones.

The Ellenberg and Mueller-Dombois (1967) physiognomic ecological classification of plant formations of the earth was obtained with a deductive method and comprises the units of formation class, formation subclass, formation group, formation, subformation, and further subdivisions. This system decidedly favours the higher units.

Often the two methods have been contrasted with each other, often because of the different interpretations of the vegetation given by different schools (individualistic, phytocoenotic, etc.), but in fact the opposition is not necessarily true: rather, it could be said that one type of approach complements the other. The vegetation typology obtained with phytosociology is very detailed, but too analytical for a real classification of the vegetation on a global level. However, this limitation is overcome by the large physiognomical categories that are favored by the deductive method.

For the classification of the vegetation of Mexico, which is very complex from the phytogeographic point of view because it developed at the contact between the Neotropical and the Holarctic kingdoms, the authors present a new system called SECLAVEMEX. They drew upon four previous proposals for the classification of the vegetation of Mexico—Miranda and Hernàndez (1963), Rzedowski (1978), COTECOTA (1994) and Gonzàlez-Medrano (2003)—for criteria, levels and nomenclature to take into consideration for the new proposal, and decided upon nine criteria: physiognomy, climate, phenology of foliage, the presence of spines and vertical spikes, leaf characteristics, succulence, floristic composition, substratum type, and secondary vegetation. The hierarchic levels were identified in great detail, as follows: bioma (4 units), major formation (15 units), formation (28 units), subformation (58 units), series of associations (134 units), association, subassociation, and facies. They also provide numerous tables that compare the SECLAVEMEX terminology with that used in the four systems of classification of the vegetation of Mexico mentioned above.

As the authors state, the SECLAVEMEX system is based on two viewpoints, first *a geographic vision with the definition of major biomes*, and second, *a botanical vision stemming from profound knowledge of species groupings*. The SECLAVEMEX system therefore comprises both higher units, the biomes and formations, and lower ones, the associations, which makes it highly innovative and original, as do the characterizations and names proposed for the various units. This explains the significance and importance of the new system of classification of the vegetation of Mexico, not only from the theoretical point of view, but also from the applicative one, above all in relation to today's grave environmental problems.

Camerino
January 2016

Franco Pedrotti

References

Bohn U, Gollub G, Hettwer C, Neuhäuslová Z, Raus T, Schlüter H, Weber H (2000) Karte der natürlichen Vegetation Europas, Maßstab 1: 2 500 000. [Map of the Natural vegetation of Europe. Scale 1: 2 500 000]. Bundesamt für Naturschutz, Bonn.

Braun-Blanquet J (1928) Pflanzensoziologie. Gründzuge der Vegetationskunde. SpringerVerlag. Berlin, Germany.

Brockmann-Jerosch H, Rubel E (1912) Die Einteilung der Pflanzengesellschaften nach okologisch-physiognomischen. Gesichtspunkten. Verlag Wilhelm Engelmann, Leipzig.

COTECOCA (1994) Revegetación and reforestación de las áreas ganaderas en las zonas áridas and semiáridas de México. Comisión Técnico Consultiva para la definición de los Coeficientes de Agostadero. México.

Ellenberg H, Mueller-Dombois D (1967) Tentative physiognomic-ecological classification of plant formations of the earth. Ber. geobot. Inst. Em, Stiftg. Rubel, Zurich., 37: 31-55.

Miranda F, Hernández XE (1963) Los tipos de vegetación de México and su clasificación. Bol. Soc. Bot. Méx. 28: 29–179.

Rzedowski J (1978) Vegetación de México. Limusa. México, D.F.

González-Medrano F (2003) Las comunidades vegetales de México.Propuesta para la unificación de la clasificación y nomenclatura de la vegetación de México. Instituto Nacional de Ecología, México, DF.

Chapter 1
Historical Survey

Introduction

Vegetation has undergone drastic changes throughout the history of man. Depending on the era, the culture, and the purposes, the approach to vegetation has been wide and varied. From the dawn of mankind, medicine men and shamans, and later philosophers and wise men, have sought to increase their knowledge of plants, their relationship with their environment, and their properties, but, above all, they have sought to learn about their potential uses. The relationship between diverse cultures and the vegetation around them has been so close that multiple plant species owe their origin to this interaction. In the same way, any number of cultures have, almost without exception, adopted as part of their ideological, mystic, religious, dietary, medicinal, and artistic customs symbols based on plant species. In principle, all learning revolved around a practical purpose, a vital necessity, but, as mankind advanced and the knowledge accumulated, the necessity of systemizing all of this information became plain, leading, therefore, to denominating plants and groups thereof according to a generally useful terminology, thus initiating the development of classification systems. Currently, the motivation to study vegetation is growing, but potential uses for plants, apart from agricultural purposes, continue to be the driving force. For example, sophisticated uses for plants are now under consideration, uses such as carbon capture, disaster mitigation or climate-condition regulation, among others.

This chapter gives a synthetic description of some historic aspects of the main schools and approaches used in the study of vegetation throughout the past, noting as well vegetation studies in Mexico. Special emphasis is placed on the course followed in times past and up to the present day by the geobotanical perspective—chorological and cartographic—in the spatial representation of vegetation. This contribution could be of assistance in terms of motivating the non-expert reader. Recommended for those interested in examining the topic in detail are Papavero et al. (1995a, b) and Romero-Zarco (2002). For the specific case of Mexico,

A. Velázquez et al., *Standardized Hierarchical Vegetation Classification*,
Geobotany Studies, DOI 10.1007/978-3-319-41222-1_1

important works include—*Vegetación de México* (Rzedowski 1978) and—*Las comunidades vegetales de México* (González-Medrano 2003).

Approaches to the Study of Vegetation

Botanical

Although various forms of ancestral knowledge have been passed down from many cultures (Mesopotamia, China, Mesoamerica and South America, among others), the oldest known ordered study of vegetation began with the Greeks to whom the term ‗botanical" owes its existence: βοτάνη (botanē)—herb. While important works by Aristotle (fourth century B.C.) and Dioscorides (first century A.D.) are, of course, known, Theophrastus (fourth century B.C.) is considered the precursor to and father of botany. Outstanding among his works is the first systematic classification for plants based on their medicinal properties, the result being that botany is perhaps the oldest branch of biology.

Later, during the Medieval Period, the Arabs stood out in the field of botanical studies. They translated and built upon the knowledge of Classical Greece and made further contributions, especially in relation to the medicinal and agronomic properties of plants (Papavero et al. 1995a, b).

From the Renaissance (fifteenth and sixteenth centuries) through to the eighteenth century, previous to Linnæus, botanical as well as other scientific knowledge grew at an unprecedented rate in comparison with the amount of advancement in the preceding centuries. In this period, navigation enabled commercial and cultural exchanges between the continents, and printing facilitated the translation and diffusion of classical and popular literature. For example, books on herbs, or —herbolaria‖, which included illustrations and descriptions of medicinal plants from different regions of the world, became popular and brought on the creation of herbaria, museums and botanical gardens as well as the use of a species nomenclature system consisting of brief Latin descriptions of the morphology of the plant. Systematic botany was developed, and numerous scientific institutions dedicated to natural history were founded. It was against this backdrop that Linnæus (Sweden 1707–1778) made his appearance. The work of Linnæus holds a special place in the history of botany, so much so that, in the year 1905 at the International Botanical Congress in Vienna, the publication date of the first edition of *Species Plantarum* (1 May 1753) was adopted as the genesis of botanical nomenclature. For over a century, Linnæus' binomial nomenclature system has been in place, but systematic botany had been drastically reoriented with the appearance of evolutionary theories.

The publication of *On the Origin of Species by Means of Natural Selection* in 1859 by Charles Darwin was a milestone surpassing the field of biology and reorienting systematic biology towards the quest for phylogenetic systems, that is the grouping of species and families not only by similarities in many of their predominating morphological characteristics, but also according to their value in reflecting kinship and analysing the evolutionary processes permitting linkages between characteristics. Since then, numerous contributions have been made to the four principal fields of botany: taxonomy, systematic, morphology and plant physiology. In the current context, in a very broad sense, botany is recognized as the study of plant species.

Plant Geography

From a different perspective towards the same object of study, plant geography was formalized with Humboldt's *Ideen zu einer Geographie der Pflanzen* in 1807, the product of his travels in South America and Mexico, in which he noted the considerable interaction between the climate, the relief, and vegetation. This vision of interaction between the biotic and the abiotic dominated the thought processes of naturalists and scientists of the day and characteristically gave equal weight to the botanical description of an area, its physical characteristics, and its spatial and temporal expression (Frey and Lösh 1998). Plant geography dominated a large part of the work of naturalists during the nineteenth century and emphasized the study of vegetation distribution expressed in taxa along physical–climatic gradients (Grisebach 1872; Werger 1974).

Geobotanical

Geobotany (plant ecology or plant geography) is the science of analysing the inseparable relationship between plants and the environment. Geobotany studies species and plant communities. From its beginnings, this discipline has conceived of the development and distribution of vegetation in close interaction with the climate, the soil, and landforms (Strasburger et al. 1982). Ancestral knowledge and the contributions of naturalists throughout the centuries along with the vision of plant geography allowed the structuring of a deeper and more integral view of vegetation called _geobotany".

The term _geobotany" was proposed in 1927 by the German Eduard Rübel and comes from the German _Geobotanik". Geobotany is the traditional name of the science concerned with the relationship between plant life and the land environment (geobiosphere). Geobotany touches upon three issues: (1) space (plant distribution), (2) the plants' habitat, and (3) changes over time to plant communities.

Over the years, geobotany has been subdivided into distinct fields of interest, which, depending on the school from which the specific subdivisions sprang and the paths then followed, have adopted different names, for example: aerobiology (also called _chorology" or vegetation cartography), coenologic geobotany (phytocoenology, botanical sociology, or vegetation science), floristic geobotany (taxonomy), ecological geobotany (processes and interactions), and historic geobotany (syngenetic or successional geobotany).

In the current context, geobotany maintains principles related to other approaches within biology (morphology, physiology, systematics, genetics, etc.) as well as with other disciplines such as geography, climatology, pedology, geology, palaeontology, and history. In the same way, geobotany has served as the conceptual base for various practical applications adopted by the agricultural and forestry sectors.

Phytosociological

At the turn of the last century, a different vision of vegetation studies, phytosociology, appeared, which concept will be expanded upon in the next chapter. The Pole Paczoski proposed the term _phytosociology” in 1896 as a denomination to summarize his conception of plant formations from the point of view of the sociability of plant species. Although this idea was widely adopted by a large part of the scientific sector in the former Soviet Union (Sukachev 1929), it remained isolated there mainly for reasons of language. It was not until the publication of the Braun–Blanquet work in 1926 developing what is known as the Zürich-Montpellier school, that the phytosociological approach took a prime position among vegetation researchers, something that continues to this day. In the twentieth century, phytosociology was the approach which best maintained the integral vision of geobotany, and, in the first quarter of the twentieth century, it was considered an approach for the study of vegetation in Europe.

In the first decades of the twentieth century, the study of vegetation in northern Europe (Scandinavian school) spawned new ideas on the morphostructural expression of vegetation; its authors, van Post, Raunkiær, and Warming, developed a system for the study of vegetation based on the life forms of plant species (cited by Schaminée et al. 1995). This system complemented elegantly the already-established phytosociological approach.

Vegetation Ecology

The term _ecology” was invented by the German zoologist Ernst Haeckel in 1869. Historically, vegetation ecology was considered synonymous with plant geography and geobotany. Nonetheless, the term _vegetation ecology” in English research gained importance owing, at a time when a large part of the scientific terminology for the study of plants was being coined, to the German term _Ökologische Pflanzengeographie”, being translated simply as _ecology” (Rübel 1927). Later, the

British vegetation research committees used the term _plant ecology" to emphasize that the research went further than floristic descriptions. Vegetation ecology currently refers to the study of space–time interactions between plant species and the environment. It is divided into autoecology (for an approach based on the study of populations of one species) and synecology (for the study of aggregates or communities of species). This synecological approach is intimately related to the concept of geobotany.

On the other hand, the Anglophone vision of plant communities was strengthened by the debate between the concepts of the —superorganism‖ proposed by Clements (1916) and that of the communities-of-individuals proposal by Gleason (1917). Both approaches have many proponents, each side disputing the other's positions for over 20 years. Many of the participants in these debates found common ground when, during the 1960s, Whittaker (1969) put forth the study of vegetation from a —continuum gradients‖ point of view. Later, on the basis of the aforementioned approaches and, exceptionally, resuming the traditional geobotanical conception, plant community ecology (phytosynecology) was consolidated as a line of research which has gained considerable momentum since the end of the twentieth century. It is known more colloquially as —vegetation science‖ which is an eclectic approach attempting to draw on the clearer and more congruent conceptual aspects of the distinct visions seen in the wide range of approaches to the rigorous study of vegetation—both of the species themselves and of their communities (Van der Maarel 1979).

Cartography of Vegetation

Corollary of Approaches Worldwide

Before the rise of specialized disciplines in particular aspects of vegetation during the last century, geobotany held an apparently marginal role in schools and universities in various parts of the world. Even so, the integrating vision of geobotany was not diminished, perhaps due to its being the approach best permitting the clearly congruent characterization, understanding, and cartographical representation of vegetation, which, in its natural setting, is presented as a mosaic. Another advantage of the integral vision of the geobotanical approach is that it established a conceptual framework to relate the physical environment to the vegetation therein, which framework is, in essence, also developed by plant science and plant ecology. Over the past few decades, the integrating vision of the geobotanical approach has gained wide currency for its importance in planning and in the management of natural resources.

Predominant Approaches in Mexico

Pre-Columbian Era

Although well-documented evidence exists (see the *1552 codex*—De la Cruz and Badiano 1940) for continuous botanical studies in pre-Hispanic times, the Tlalocan mural, dating from 300 B.C. located in Tepantitla, Teotihuacan, is the oldest pictorial source for medicinal herbalism in Mesoamerica. That said, it is a fact that most material evidence for pre-Hispanic traditional knowledge on the study of vegetation has been lost. Fortunately, communities surviving the conquest preserved, enriched, and transmitted a goodly part of their traditional knowledge to the present generation (Gómez-Pompa 1993).

During the pre-Columbian era, the amount of advancement in the scientific study of vegetation in Mesoamerica and the Andes region was significantly greater than in Europe. For example, chroniclers noted the existence of numerous botanical and zoological gardens, and there were very advanced plant and animal classifications including information on their uses, properties, and their life forms and on the characteristics of their habitat. In this way, in many senses, ethnobotany has recognized the existence of a true pre-Hispanic botanical science (Gomez-Pompa 1993).

Colonial Era Through the Beginning of the Twentieth Century

The first scientific report published on American plants was made by the medical doctor Nicholás Monardes from Seville, who, in 1565, wrote the first treatise on medicinal plants in New Spain (Monardes 1565; Boxer 1963). What is interesting about this work is that its author never travelled to New Spain, but rather he described plants brought back to Spain, which he then cultivated in a dedicated botanical garden (Somolinos 1971). At this time, vegetation researchers who did not cross the Atlantic obtained information on plants from ships' captains, missionaries, and travellers (Emmart 1940). In Monardes' treatise, plants are described for the first time as _allspice plant" (*Pimienta dioica*), _golden shower tree" (*Cassia fistula*), _Jalap bindweed" (*Ipomoea jalapa*), tobacco (*Nicotiana tabacum*), sassafras (*Sassafras albidum*), pineapple (*Ananas sativus*), guava (*Psidium guajava*), and many others.

Due to the excellent descriptions of these new varieties and plant species, Monardes is considered one of the most notable precursors of American pharmacognosy (Gómez-Pompa 1993). Perhaps, however, the most notable work on American medicinal plants was that of the indigenous doctor Martín de la Cruz, *—Manuscrito o Código Badiano*‖ (Emmart 1940), which was translated by Juan Badiano, a native of Xochimilco. Much of the knowledge on plants in Mexico's colonial era was based on their uses and properties. That said, there was also formal botanical research dating to the end of the sixteenth century, notably that of

CAROLI LINNÆI
S:Æ R:GIÆ M:TIS SVECIÆ ARCHIATRI; MEDIC. & BOTAN. PROFESS. UPSAL; EQUITIS AUR. DE STELLA POLARI; nec non ACAD. IMPER. MONSPEL. BEROL. TOLOS. UPSAL. STOCKH. SOC. & PARIS. CORESP.
SPECIES PLANTARUM,
EXHIBENTES
PLANTAS RITE COGNITAS,
AD
GENERA RELATAS,
CUM
DIFFERENTIIS SPECIFICIS,
NOMINIBUS TRIVIALIBUS,
SYNONYMIS SELECTIS,
LOCIS NATALIBUS,
SECUNDUM
SYSTEMA SEXUALE
DIGESTAS.
TOMUS I.
Cum Privilegio S. R. M:tis Sueciæ & S. R. M:tis Poloniæ ac Electoris Saxon.
HOLMIÆ,
IMPENSIS LAURENTII SALVII.
1753.

Fig. 1.1 Front page of the Species Plantarum contribution

Francisco Hernández, in which an affinity for geobotany's integrating concept can be seen. Although nothing of note occurred in Mexican botanical studies for almost 100 years after Hernández, European botanical research underwent a revolution with the publication in 1753 of Swedish naturalist Carl Linnæus' two-volume —*Species Plantarum*‖ (Fig. 1.1).

In the eighteenth and nineteenth centuries, there were interesting botanical works such as those of Vicente Fernández, Martín Sessé, and Luis Martín, as later recognized by Alexander Humboldt in his 1807 essay on plant geography. Since then, Mexico's great geobotanical diversity has been evident, thanks to abundant references in Humboldt's works. In the nineteenth century, geobotanical research in Mexico was carried out in close cooperation with geologists, chemists, and physicists, given that vegetation is an indicator of Mexico's complex mineral wealth, of use in locating new veins of various minerals and in the discovery of different chemical compounds. Martens and Galeotti drew up the first geobotanical classification proposal based on climatic–geographic input in 1842 (González-Medrano 2003). Later, in 1869, Herrera published his article —*Apuntes para la geografía botánica de México*‖, in which he recapitulated the importance of the physical component to the understanding of plant distribution. Some years later, Grisebach

(1872), who collaborated closely with Humboldt and who is today considered the father of geobotany, presented his work, —*Végétation du globe*‖, in which the strong correlation between the climatic, geomorphologic, and edaphic aspects of the shaping of plant formations is highlighted. This work includes a section on Mexico illustrating in detail a first approach to the alpine vegetation in the Trans-Mexican Volcanic Belt. Only three years later came Hemsley's Contribution, —*Bosquejo de la Geografía y rasgos prominentes de la flora de México*‖ (González-Medrano 2003), in which the floristic section is a compilation of many earlier geobotanists' observations demonstrating the great richness of plant life in México.

In 1890, Oscar Drude's —*Handbuch der Pflanzengeographie*‖ came out. It deals with the geobotany of central Mexico. Later, in 1899, Ramírez published his book, —*La vegetación de México*‖, the first of its kind, which was based on his own observations and which noted geographic–botanical co-regions. With the exception of Humboldt, Martens, and Galeotti, a large amount of the research published was carried out from compilations of previous works without any direct knowledge of the country.

The next important step in Mexican ethnobotany came some time after independence with the creation of the Instituto Médico Nacional (IMN) in 1890, the mission of which was the scientific study of medicinal plants. This prestigious institution had a long and brilliant scientific trajectory as witnessed by the numerous publications appearing in learnèd journals of the day (*La naturaleza* and *Anales del Instituto Médico Nacional*) before its closure in 1915. Among its collaborating researchers were Fernando Altamirano, author of various publications on medical plants, and Alfonso Herrera, one of the most notable biologists in the history of Mexico. The IMN established the most formal herbarium then in existence, which body served as the forerunner to the current Herbario Nacional.

During the first revolutionary period, the IMN was transformed into the Dirección de Estudios Biológicos (DEB) within the Secretaría de Fomento, Colonización e Industria. This secretariat also includes the Museo Nacional de Historia Natural and the Comisión Geográfico Exploradora (CGE), headed by Alfonso Herrera. Among other researchers collaborating with the CGE can be found notable botanists such as Maximino Martínez, Caziano Conzatti, and Isaac Ochtorena (Beltrán 1977). Without a doubt, this was the beginning of a brilliant era of expansion within the field of biology in Mexico, given that the first marine biology station, the zoological park, and the botanical garden at Chapultepec were established in addition to the first conservation movement.

The year 1929 saw the DEB come to an end, with its installations and collections being passed to the Instituto de Biología de la Universidad Nacional Autónoma de México. The following years are difficult to detail. Numerous botanical projects were set up across various institutions, each often addressing different aspects of the field, until, in 1962, the Sociedad Botánica de México (SBM) was created. It is difficult to recognize any school of vegetation in it, given that there have been contributions to differing fields relating to the study of vegetation from both Mexican and foreign specialists. There is no doubt, however, that one notable contribution, for the purposes of this work, was the Mexican vegetation classification system proposed by Miranda and Hernández-X in 1963.

Contemporary Period

A long list of geobotanists appeared after the work of Ramírez, among whom stand out Fournier 1900 (quoted by Gonzalez-Medrano 2003; Harshberger 1911; Ochoterena 1918, and Sanders 1921). Although there were many other efforts made regionally to study vegetation, these authors had already presented the first formal maps of the geobotanical regions of Mexico. Ochoterena (1937) made a transcendent contribution to the subject, giving a detailed description of the geographic areas of plant distribution across Mexico in conjunction with rigorous and organized floristic lists in a hierarchical classification system running from formations to —sinucial‖. This contribution inspired Contreras (1941) to shape his own on plant symorphy in Mexico in which he produced a map drawing on the climatic regions designated by various authors. From that time onward, the cartographic representation of vegetation becomes a core theme, resulting in various authors presenting proposals on the biotic provinces of Mexico, among the most notable being the works of Dice (1943), Smith and Johnston (1945), Goldman and Moore (1946), and Leopold (1950). Later, Macías-Arellano (1952) published a translation of Leopold's —*Formas de vegetación de México*‖. Almost at the same time, Aubreville (1962) alongside Miranda and Hernández-X (1963) published complementary works, though based on distinct approaches. The first consists of a redefinition of the plant formations of Mexico inserted into the global classification system of the era, while Miranda and Hernández-X published —*Clave para identificar los tipos de vegetación de México*‖, which presents information on the vegetation of Mexico with descriptions, photographs, dominant species, and climatic data. Since then, a long list of methods for understanding the vegetation of Mexico has been presented, between them profiling the visions and groups dedicated to the floristic–botanical field and ecology as well as to the sylvan and geographic facets (cartography). In this last aspect, the vegetation classification system occupied a central place and began to include cartographic legends. One example is the —*Mapa de tipos de vegetación de la República Mexicana*‖ by Flores Mata et al. (1971), which, at a scale of 1:2,000,000, differentiates between 25 vegetation types.

In recent times, there have been notable advances in the study, knowledge, and cartographic representation of the vegetation of Mexico. One of the most important contributions to knowledge is found in Rzedowski's —*Vegetación de México*‖ (1978), in which there appears a vegetation map of the entire country. This contribution, along with that of Miranda and Hernández-X (1963), has been the most widely accepted for use in Mexico. Attention should also be drawn to the major role played by the *Bóletin* of the SBM in which, over the decades, various studies have been published, diffusing the knowledge they have generated on the vegetation of Mexico and of other parts of Latin America. Equally, various research institutes have contributed with the publication of lists of local floristics and flora (Lott 1993; Espinosa and Rodríguez 1995a, b; Rodríguez and Espinosa 1995, 1996a, b; Padilla-Velarde et al. 2006 among many others). At the same time, from the governmental sector, the Comisión nacional para el conocimiento y uso de la biodiversidad (CONABIO) through the Sistema Nacional de Información sobre Biodiversidad (SNIB) has lent support to the integration of a digital database of plant collections, putting Mexico in an unprecedented position in terms of knowledge on phytodiversity.

CONABIO during 2008 was able to compile a series entitled: "The Natural Capital of Mexico". In it, Challenger and Soberon (2008) were able to provide a synthesis of the vegetation types mostly used by botanist which served to build upon map legends made by the official Mexican mapping agency (INEGI).

For their part, in academic circles, many of the higher educational institutions in Mexico have developed lines of research into the ecology of plant populations and communities. These schools are dominated by the Anglo-American approach and, commonly, vegetation mapping is used only marginally as a general reference and study-site locator.

Unlike in South America (Rangel-Ch. et al. 1996, 1999, 2002; Navarro and Maldonado 2002), the phytoecociological approach has been applied less in Mexico, and, at that, mainly in the northeast. That said, it is foreseen that its use will increase markedly because: (1) of Mexico's increasing taxonomical knowledge, which is fundamental to the rigorous development of phytosociology, (2) of the relevé method and a standard classification system which will help generate information across the country in the short term and will also facilitate monitoring, (3) other highly diverse Latin American countries, such as Colombia and Bolivia, have demonstrated its potential use, (4) the phytosociological approach is being adopted in vegetation studies and classification systems in Canada and the United States (Jennings et al. 2002; Snow 2004), and (5) conceptually speaking, phytosociology creates the necessity to amass precise information, derived from vegetation studies with units of vegetation able to be represented in maps, thus potentiating vegetation cartography.

References

Aubreville A (1962) Temas fitogeográficos: 1. Clasificación fisonómica and florística de las formaciones vegetales tropicales: 2. Clasificación ecológica de las formaciones vegetales tropicales, 3. Clasificación de las principales formaciones vegetales de México. Inst. Mex. Rec. Nat. Renov. A.C. México, D. F.

Beltrán E (1977) Medio siglo de recuerdos de un biólogo mexicano. Soc. Mex. Hist. Nat. México, D. F.

Boxer CR (1963) Two pioneers of tropical medicine: García d'Orta and Nicolás Monardes, Wellcome Historical Medical Library, London.

Braun-Blanquet J (1926) Études phytosociologiques en Auvergne. Impr. de G. Mont-Louis, Clermont-Ferrand.

Challenger A. Soberón J (2008) Los ecosistemas terrestres. In: Capital Natural de México, Vol. I: Conocimiento actual de la biodiversidad. CONABIO, Mexico City: 87–108.

Clements FE (1916) Plant succession: an analysis of the development of vegetation. Carnegie Inst. Wash. Publ.

Contreras A (1941) Ensayo de localización de las simorfias vegetales dominantes en la República Mexicana. Secretaría de Agricultura and Fomento. Tacubaya, D. F.

De la Cruz M, Badianus J (1940) The Badianus manuscript –an Aztec herbal. Baltimore, Md.
Dice IR (1943) The Biotic Provinces of North America. Ann Arbor. Michigan.
Emmart EW (1940) The Badianus Manuscript (Codex Barberini, Latin 241). Vatican Library; an aztec herbal of 1552. Introduction, translation and annotations by E. W. Emmart. The Johns Hopkins Press, Baltimore.
Espinosa G, Rodríguez LS (1995a) Listado florístico del estado de Michoacán. Sección IV (Angiospermae: Fagaceae, Gramineae, Krameriaceae, Leguminosae). Flora del Bajío and de Regiones Adyacentes. Fascículo complementario XII. Instituto de Ecología A.C. Centro Regional del Bajío. Pátzcuaro, Michoacán. México.
Espinosa G, Rodríguez LS (1995b) Listado florístico del estado de Michoacán. Sección II (Angiospermae: Compositae). Flora del Bajío and de Regiones Adyacentes. Fascículo complementario VII. Instituto de Ecología A.C. Centro Regional del Bajío. Pátzcuaro, Michoacán. México.
Flores-Mata G, Jiménez L, Madrigal S, Moncayo R., Takaki T (1971) Tipos de Vegetación de la República Mexicana. Secretaría de Recursos Hidraúlicos. México, D.F.
Fournier EPN (1900) Mexicanas Plantas Nuper a Collectoribus Expeditionis Scientificae Allatas aut Longis ab Annis in Herbario Musei Parisiensis Depositas Praeside J. Decaisne... Enumerandas Curavit. Parisiis : ex Typographeo Reipublicae, 1872-1886.
Frey W, Lösh E (1998) Lehrbuch der Geobotanik. Pflanze und Vegetation in Raum und Zeit. Gustav Fisher, Stuttgart.
Gleason HA (1917) The structure and development of the plant association. Bull. Torrey Botanical Club 43: 463- 481.
Goldman EA, Moore RT (1946) Biotic provinces of Mexico. Journal of Mammalogy 26: 347–360.
Gómez-Pompa A (1993) Las raíces de la etnobotánica mexicana In: Guevara S, Moreno-Casasola P & Zedowski J (compiladores). Logros and Perspectivas del Conocimiento de los Recursos Vegetales de México en vísperas del Siglo XXI. Instituto de Ecología A.C. and Sociedad Botánica de México: 26–37.
González-Medrano F (2003). Las comunidades vegetales de México.Propuesta para la unificación de la clasificación y nomenclatura de la vegetación de México. Instituto Nacional de Ecología, México, DF.
Grisebach AHR (1872) Die Vegetation der Erde nach ihrer klimatschen Anordnurg. Leipzig, Engelmann.
Harshberger JW (1911) Phytogeographic Survey of North America. G. E. Stechert, New York.
Jennings M, Loucks O, Glenn-Lewin D, Peet R, Faber-Langendoen D, Grossman D, Damman A, Barbour M, Pfister R, Walker M, Talbot S, Walker J, Hartshorn G, Waggoner G, Abrams M, Hill A, Roberts D, Tart D. (2002) Standards for associations and alliances of the U.S. national vegetation classification. Version 1.0. Vegetation Classification Panel, Ecological Society of America. Washington, D.C. USA.
Leopold AS (1950) Vegetation zones of Mexico. Ecology. 31: 507–518 (1735-1770).
Lott EJ (1993) Annotated checklist of the vascular flora of the Chamela Bay region, Jalisco, Mexico. Occas. Pap. Calif. Acad. Sci., 148: 1-60
Macías Arellano L (1952) The Future of Mexico Lies in Education, In: Transactions of the Fifteenth North American Wildlife Conference, 9–10.
Miranda F, Hernández XE (1963) Los tipos de vegetación de México and su clasificación. Bol. Soc. Bot. Méx. 28: 29–179.
Monardes N (1565*) Historia medicinal de las cosas que se traen de nuestras Indias Occidentale* Dos libros, el uno que trata de todas las cosas que se traen de nuestras indias occidentales, que sirven al uso de medicina and como se ha de usar la raíz de Mechoacan purga excelentisima and el otro que trata de piedra de bezaar and de la yerba escorzonera. En casa de Alfonso Escrivano, Natural, Sevilla.
Navarro G., Maldonado M (2002) Geografía Ecológica de Bolivia. Vegetación and Ambientes Acuáticos. Centro de Ecología Simón I. Patiño, Cochabamba.
Ochoterena (1918). Las regiones geográfico-botánicas de México. Bol. Soc. Mex. Geog. Est. V(8): 221–231.

Ochoterena I (1937) Esquemas biotípicos and Sinecias características de las regiones geográfico-botánicas de México. An. Inst. Biol. Mex. 8: 463–597.
Padilla-Velarde, Cuevas-Guzmán ER, Ibarra-Manríquez G, Moreno-Gómez S (2006) Riqueza and biogeografía de la flora arbórea del estado de Colima, México. Revista Mexicana de Biodiversidad. México, D. F. 77: 271–295.
Papavero N, Llorente J, Espinosa-Organista D (1995a) Historia de la Biología comparada. Desde el génesis hasta el siglo de las Luces. Vol. I, UNAM, México, D.F.
Papavero N, Scrocchi GJ, Llorente J (1995b) Historia de la Biología comparada. La Edad Media. Vol. II, UNAM, México, D.F.
Rangel-Ch, J.O, Lowy-C P, Aguilar-P M, Garzón-C A, Ruiz-C P, Castaño M O (1996) Inventario de la Biodiversidad en Colombia. Cespedesia 21(67): 511–514.
Rangel C., J. O., D. Sánchez & C. Ariza (1999) Fitosociología del páramo de Frontino. In: C.A., Velásquez-R., Parra LN, D. Sánchez., Rangel-Ch JO, Ariza C & Jaramilo A. Tardiglacial and holoceno del norte de la cordillera Occidental de Colombia: 110–120 (Más anexos). Universidad Nacional de Colombia- COLCIENCIAS, Medellín.
Rangel JO, Bogotá-A & Jiménez-B (2002) Atlas palinológico de las familias mas diversificadas en la Amazonia Colombiana. In: Rangel-Ch JO, Aguirre-C J & Andrade-Correa MG (eds), Libro de Resúmenes Octavo Congreso Latinoamericano y Segundo Colombiano de Botánica. Instituto de Ciencias Naturales, Universidad Nacional de Colombia, Bogotá.
Rodríguez JLS, Espinosa GJ (1995) Listado florístico del estado de Michoacán. Sección I (Gymnospermae; Angiospermae: Acanthaceae-Commelinaceae). Flora del Bajío and de Regiones Adyacentes. Fascículo complementario VI. Instituto de Ecología A.C. Centro Regional del Bajío. Pátzcuaro, Michoacán. México.
Rodríguez JLS, Espinosa GJ (1996a) Listado florístico del estado de Michoacán. Sección III. (Angiospermae: Connaraceae-Myrtaceae excepto Fagaceae, Gramineae, Krameriaceae and Leguminosae). Flora del Bajío and de Regiones Adyacentes. Fascículo complementario X. Instituto de Ecología A.C. Centro Regional del Bajío. Pátzcuaro, Michoacán. México.
Rodríguez JLS, Espinosa GJ (1996b) Listado florístico del estado de Michoacán. Sección V. (Angiospermae: Najadaceae-Zygophyllaceae). Flora del Bajío and de Regiones Adyacentes. Fascículo complementario XV. Instituto de Ecología A.C. Centro Regional del Bajío. Pátzcuaro, Michoacán. México.
Romero-Zarco C (2002) Historia de la Botánica. Apuntes de Biología. Universidad de Sevilla, España. (http://www.aloj.us.es/optico/carromzar/botanica1/T1historia.htm).
Rübel E (1927) Ecology, plant geography, and geobotany; their history and aim. The Botanical Gazette 84: 428–438.
Rzedowski J (1978) Vegetación de México. Limusa. México, D.F.
Sanders EM (1921) The natural regions of México. Geog. Rev. 11: 212–226.
Schaminée JHJ, Weeda EJ, Westhoff V (1995) De vegetatie van Nederland. Deel 2. Plantengemeenschappen van wateren, moerassen en natte heiden. Opulus Press, Uppsala-Leiden.
Smith AC, Johnston IM (1945) A phytogeografic sketch of Latin America. Plants and plant science in Latin America, 11–18.
Snow K (2004) HierarchySort. (ORGANIZACIÓN ABI) IVC Hierachy full list.
Somolinos D'Ardois G (1971) El Doctor Francisco Hernández and la Primera Exploración Científica de América. SEP-Setentas. México.
Strasburger E, Noll F, Schenck H, Schimper AFW (1982) Tratado de Botánica. Omega, 7ª ed., Barcelona.
Sukachev VN (1929) Über einige Grundbegriffe in der Phytosoziologie. Ber. Deutsch. Bot. Gesell. 47: 296–312.
Van der Maarel E (1979) Transformation of cover-abundance values in phytosociology and its effects on community similarity. Vegetatio 39: 97–114.
Whittaker RH (1969). Evolution of diversity in plant communities. In "Diversity and Stability in Ecological Systems". Brookheaven Symp. Biol. 22. 178-196. Springfield, Va. National Bureau of Standards, U.S. Dept. of Commerce.

Chapter 2
Background on Vegetation Classification Systems

Introduction

Vegetation is formed by plant species growing as the result of a long development process consistent with the places they inhabit and which constitutes a fundamental part of ecosystems. As a result, vegetation is understood as resulting from the effects of various environmental factors on sets of species which cohabit in a determined space and period (Velázquez and Romero 1999; FGDC 2005). Generally speaking, two major categories can be seen: cultural vegetation and natural vegetation. The former owes its development, permanence, and expression to processes regulated by the continuous activity of mankind, for example agricultural areas and planted forests, which are a relatively recent expression (to the order of decades and centuries). Natural vegetation, by contrast, is underlain by processes in the main determined by climatic, geomorphological, edaphic, and ecological components, which interact simultaneously and continuously over extended periods of time (to the order of hundreds to thousands of years). As the effects of human actions have become more evident, it has become increasingly difficult to determine the limits between natural and cultural vegetation, though international agreements exist which allow for the understanding of irrefutable predominance. Profound knowledge of vegetation in general, the directional forces determining its presence notwithstanding, is fundamental for multiple purposes as much locally as globally, particularly in terms of better vegetation management and conservation (Matteucci and Colma 1982; FGDC 2005).

This chapter serves as a brief description of the diversity and approaches fundamental to various vegetation classification systems employed at the global level and forming the backdrop against which the main expressions used in Mexico and the systems most highly regarded and employed in research and in the management of natural resources in the country are described. It should be mentioned that there exist different vegetation classification system proposals for Mexico, and the first

A. Velázquez et al., *Standardized Hierarchical Vegetation Classification*,
Geobotany Studies, DOI 10.1007/978-3-319-41222-1_2

question is whether, in reality, it was necessary to write this chapter. The arguments for its inclusion in this present work are explained in the following paragraphs.

The abundant terminology existing in vegetation research materials reflects, in part, the physiognomic–structural–floristic complexity of vegetation communities found in Mexico. Moreover, there is interest in the need to recognize the heterogeneity of the vegetation and of its typification through the names characterizing it. In general, the prevalent initiatives have been those concerned with local demands, from which, when the need arises to extrapolate to the macroregional and national levels, conflict arises from the lack of compatibility in the criteria used in creating classifications and nomenclature. Equally, confusion is generated by a plethora of recurring synonyms. Faced with such circumstances, it is difficult or well-nigh impossible to use local information, as plentiful as it may be, to integrate data for analysis at the national level or even to make comparisons between regions sharing common characteristics (Huber and Riina 2003). The proposals covering the most territory and most commonly used among academics are not free of inconsistencies and synonyms, thus contributing to the prevailing entropy surrounding the Mexican vegetation classification system (González-Medrano 2003). Despite multiple efforts, there is still no generally accepted compatible, hierarchical standardized system in Mexico.

Research into plant species populations (autophytoecology) can dispense with relying on a single standardized, hierarchical system for vegetation in Mexico. On the contrary, for research on vegetation communities (synphytoecology), it is fundamental to identify the unit of research, and it is here where terms such as —vegetation type‖, —plant community‖, and —vegetation association‖ are used time and again with different connotations and generally applied in different ways across various regions and between academics, resource managers, and local data collectors. Currently, with the boom in remote sensing sources and the availability of spatially explicit databases handling geographic information systems, synphytoecological studies have become even more complex. On the one hand, no clear unit of research applies from the purely vegetational point of view, where plant species aggregates are the —unit‖, whereas from the purely spatially explicit point of view, land cover classes are solely recognized as the —unit‖ (Velázquez et al. 2006). This ambiguity makes for difficulties when it comes to planning systemized research, natural resource management, or conservation strategic planning and work, in fact, ultimately, in taking any action within the scope of conserving environmental goods and services (Carabias and Landa 2006). Global, national and local policy-makers dealing with natural resource management need to be provided with sound, irrefutable units describing vegetation as a basis for policy-making.

Multiple examples illustrate the necessity of relying on a single classification system in Mexico, as in most countries worldwide and on its corresponding syntaxonomy and nomenclature. For the study of vegetation sensu stricto, it is fundamental to understand its spatial distribution and to derive conclusions from it, which conclusions can then be applied beyond the sample conditions. This task is currently impossible in Mexico given that the majority of vegetation studies, with very few exceptions, do not present complete floristic lists, whereas the remaining show the preserve of taxonomic interest groups or are limited by the accessibility of the sampling area (Bojórquez-Tapia et al. 1995). For a greatly diverse country possessing such a geological heritage, it would be desirable to rely on databases which include, taxonomically speaking, a high proportion of all species within the country. Examples of systematic vegetation databases in general permit descriptions of habitat conditions and of entire continents such as Europe or complete countries of enormous territorial proportion and diversity such as Canada and Colombia, which prioritize their research needs based on —spaces‖ or —types of vegetation‖ not yet thoroughly researched (http://europa.eu/scadplus/leg/nl/vb/128076.html).

The best-known defenders of natural vegetation are mostly from the biological sciences guild (ecologists, agronomist, foresters, and geographers, among others). Mexico is no exception, and, in all cases, the ecosystemic vision predominates as the framework for sustainability. The great challenge lies in the lack of a single inventory of ecosystems in Mexico. Thus, vegetation tends to be fallen back on as a substitute. At the same time, the territorial expression of ecosystems is subordinated to plant cover. The vegetation as much as the land cover, however, lacks a rigorous inventory. For example, in the following table, a comparison is made over a 24-year period. The loss or gain value for a category of plant cover therefore depends on its definition and eventually on its territorial expression, a task in which a single classification system would be of enormous assistance in terms of reducing incongruence (Table 2.1).

Each block shows that there are different data and demonstrates the importance of relying on a vegetation classification system which permits losses and gains in each category to be measured (Velázquez et al. 2002).

In contrast to common practice across most of Europe, where the phytosociological classification system has been widely adopted in naming most plant communities, on the American continent the lack of consensus on establishing one or various vegetation classification systems is notable. Over a decade ago, a group of vegetation researchers set out to create an international classification of ecological communities for the terrestrial vegetation of Canada and the United States, and now, more than 10 years later, the creation of an international vegetation

Table 2.1 The area covered by a plant formation measured at two different times allows the rate of change and net loss per year to be calculated

Plant formation	t^1 (1976) km^2	t^3 (2000) km^2	Rate r	Net (Ha/year)
Forest	352,117	327,810	−0.003	−40,708
Jungles	378,367	307,207	−0.009	−111,106
Shrubs	604,933	557,089	−0.003	−79,687
Forest and jungles	730,484	635,027	−0.006	−154,391
Woody vegetation	1,335,417	962,422	−0.014	−547,288

If plant formation aggregation changes, so it does the rate, hence precise definition on what to compare with what revels critical for understanding vegetation trends (*Source* Velázquez et al. 2010a)

classification panel has been initiated on the continent of which Mexico forms part. The proposal put forward by this commission is not to create a unified vegetation classification system, but rather to propose systems which maintain a consensual platform with the aim of facilitating communication as well as research on the subject (FGDC 2008).

In Mexico, despite the information amassed on the characteristics and composition of different vegetation units from the local through regional levels, there is still not a hierarchical and standardized vegetation classification system, which is consistent with the criteria and levels seen in North America and in the rest of the world and which includes the benefits of previous proposals, such that it might become widely used both by academics and by those responsible for the management of natural resources. In this chapter, background on various classification systems and approaches to classification having affected Mexico are presented. In the following chapters, a proposal outlining the fundamentals of a better classification system for vegetation types in Mexico is developed in detail. It aims for the production of maps and legends detailing plant cover on different scales, for the deepening of knowledge on the plants of Mexico and the generation of information leading to the creation of botanical research strategies applied to the better management and conservation of natural resources.

Broad-Scope Vision Classification Systems

Throughout the centuries, the study of vegetation classification has been an ongoing activity. To speak of vegetation classification systems means describing as many classifications as there are authors on the subject. Arising from diverse contexts, many approaches and schools have sprung up, all working towards the same goal—the development of vegetation classification systems—and all born as a response to the problems presented by different regions of the planet (Whittaker 1962). In this context, Huber and Riina (2003), for example, cite 3,500 traditional phytoecological terms used in Mexico, Central America, and on the islands of the Caribbean, showing the breadth of the reference framework generated by the physiognomic–structural–floristic complexity of plant communities across the region.

Among the first plant community classifications is found Humboldt's —*Hauptformen*‖ (1806) used by Grisebach 30 years later in the sense of —Formations‖ to denote physiognomic characteristics in vegetation. Originally, these formations made reference to a uniformity in their dominant biological spectra and made it possible to make comparisons among distinct biogeographical regions. A century later, at the third International Botanical Congress held at Brussels in 1910, the concept of —plant formation‖ was proposed as an expression of determined conditions for life (Table 2.2).

For their part, Holdridge et al. (1971) propose that —life zones‖ on the planet can be classified through the analysis of vegetation and from some basic climatic data—that is to say that the physiographic conditions of the land, the complexities

Table 2.2 Correspondence between distinct climatic and bioclimatic criteria determining the principal plant formations globally

Plant group (De Candolle)	Postulated plant requirements	Formation	Köppen climate class
Megatherms	Continuous high temperature and abundant moisture	Tropical rain forest	A (tropical rainy climates)
Xerophiles	Drought and heat tolerant	Tropical desert	B (dry climates)
Mesotherms	Moderate temperature and moisture	Temperate deciduous forest	C (warm temperate climates)
Microtherms	Less heat and moisture, tolerate long cold winters	Boreal forest	D (snowy-forest climates)
Hekistotherms	Tolerant of conditions beyond the tree line	Tundra	E (polar climates)

Translated from Archibold (1995: 2)

of the soil and the plant communities together with the animal component are fundamentally subordinated to the macroclimate. This method divides the earth's macroclimates into over 100 ecologically equivalent units, with each unit representing a sector of the climatic continuum and hosting a certain set of plant associations (Fig. 2.1).

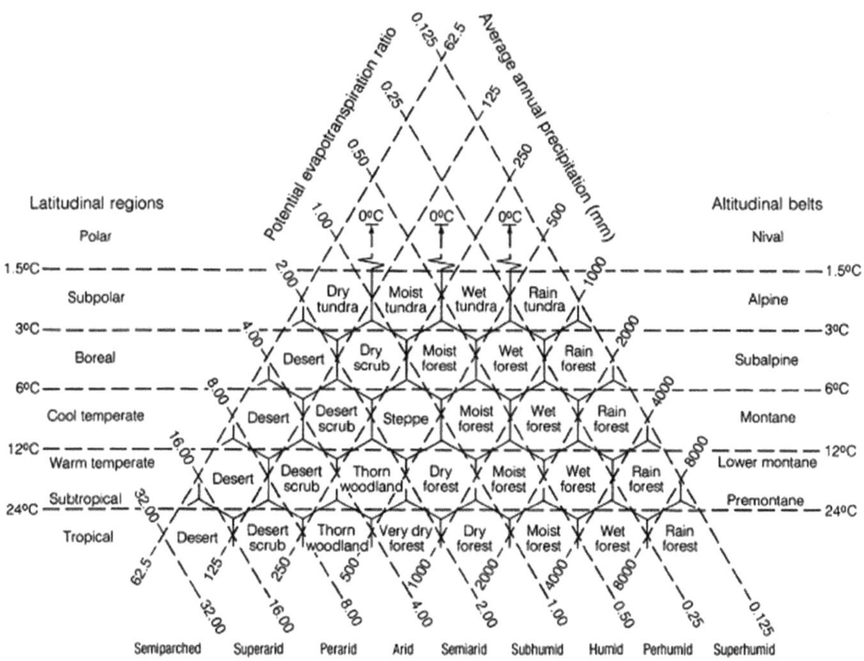

Fig. 2.1 Diagram of life-zone classifications or world plant formations according to Holdridge (1967) (taken from Archibold 1995: 3)

Table 2.3 General summary of the thermal limits of the principal types of climate in Mexico (García 2004)

Humid climates	Symbols	Mean annual temperature
Hot	*A*	Above 22 °C
Semi-warm	*A(C); (A)C*	Between 18 and 22 °C; above 18 °C
Temperate	*Ca; Cb*	Between 12 and 18 °C
Semi-cold	*C(b') Cc*	Between 5 and 12 °C
Cold	*E(T) C*	Between −2 and 5 °C
Very cold	*EF*	Below −2 °C
Arid climates	Symbols	Mean annual temperature
Hot	*B(h')*	Above 22 °C
Semi-warm	*Bh'(h); Bh*	Between 18 and 22 °C
Temperate	*Bk; Bk'*	Between 12 and 18 °C
Semi-cold	*B(k")*	Between 5 and 12 °C

From the climatic perspective, García (2004) lays out for Mexico a series of modifications to Köppen's classification referring to temperature (Table 2.3), in which she identifies two large groups—humid and arid. The first group is divided into five thermal categories, with the second divided into four. The use of climatic information has been constant as much for defining plant formations as for their typification and nomenclature.

The first plant classifications from the floristic approaches applied at a global level were based on —floristic kingdoms‖, taking into account the differences between the continents as per their particular formation. In this sense, Good (1953) defined six floristic empires or kingdoms in the following manner:

(I) Boreal (North Temperate Zone)
(II) Paleotropical (tropical zones in the Old World) subdivided into: (a) *African*, (b) *Indomalayan,* and (c) Polynesian
(III) Neotropical (tropics in the Americas)
(IV) Southern Africa (with the exception of the Cape region of South Africa, which is home to exceptional flora)
(V) Australian (the most isolated)
(VI) Antarctic (including the South African cape and islands such as those of New Zealand).

Riley and Young's classic work (1966) also defined from an ecogeographic perspective eight types of plant formations: (1) *deciduous forests*, (2) *coniferous forests*, (3) *rain forests*, (4) *grassland*, (5) *savannas*, (6) *arid zone vegetation*, (7) *tundra and alpine vegetation,* and (8) *coastal and freshwater vegetation*. Using the same approach, Takhtajan (1986) establishes floristic kingdoms, regions, provinces, sectors, districts, and tesserae. Later, Dinerstein et al. (1995) divided the world into five ecosystems and 200 ecoregions further divided into distinct ecodistricts and ecosections.

Table 2.4 Classes and leaf measurements from Raunkiær (Raunkiær's modification 1934)

Class	Designation	Size (mm^2)	Measurement proposed for Mexico (mm^2)
1	Leptophyll	0–25	0–25
2	Nanophyll	25–225	25–125
3	Microphyll	225–2,025	125–625
4	Mesophyll	2025–18,225	625–3,125
5	Macrophyll	18,225–164,025	3,125–15,625
6	Megaphyll	>164,025	>15,625

Olson et al. (2001) proposed 867 terrestrial ecoregions, classified into 14 different biomes.

Physiognomic approaches to the classification of plant communities are developed by emphasizing the external appearance of the plants in their totality to indicate an adaptive connotation. A predecessor to many of these proposals was Raunkiær's —*Life forms*‖ (1934), which was based on the bud position and established five main categories (phanerophytic, camephytic, hemicryptophytic, cryptophytic, and terophytic), which permitted comparisons of the plant composition differentials in the major ecosystems of the world, and which also, in an attempt to facilitate statistical research on the communities, established six classes of leaf size (Table 2.4), relating longitude and leaf area to the climate even where the structure of the floristic composition differed widely (see Granados and Tapia 1990).

The specialization processes seen throughout the nineteenth and twentieth centuries motivated the development of proposals, which, though more specialized, were based on data from particular regions and localities. This approach gave rise to a wide range of possibilities or approaches described herein below.

Approaches to Vegetation Classification

According to Fosberg (1961), vegetation is defined by its external appearance (physiognomic approach) and is the result of the Bioma structure (structural approach), the proportion of life forms (functional approach) and the coarse features of the floristic composition (floristic approach). There exists the eventual possibility of integrating these criteria to define holistic classification approaches from a landscape angle. In this way, according to Granados and Tapia (1990), the main approaches to vegetation classification prevalent in the world and, by consequence, in Mexico can be grouped as (*a*) physiognomic, (*b*) structural, (*c*) functional, (*d*) floristic, and (*e*) landscape.

Physiognomic

Physiognomic characteristics have formed the base for many vegetation classification systems, that is their physical characteristics and their external appearances (e.g. tall perennial forests, low deciduous jungle, shrublands), and have been widely used for both local and regional studies. They can refer both to growth forms of the dominant species and to leaf characteristics (seasonality, shape, phenology, duration, size, and texture). These attributes provide a quick and efficient mechanism for characterizing vegetation (González-Medrano 2003).

There exist as many vegetation classification systems as researchers and conditions, and, as an example, there is a refined school which includes measurements in a determined area as to the dominance detected through three elements: number, distribution, and biomass of the individual plants (Granados and Tapia 1990). Among methods for the physiognomic description of vegetation, those of Cain (1932) and Cain and Castro (1959) are widely used. They distinguish between two types of dominance: (1) physiological or ecological and (2) physiognomic. Küchler's system (1949) uses parameters such as easily observed life forms, measurements, and representation and, are, moreover, applicable to any region, making them useful in comparative studies. Dansereau's physiognomic system (1951) relates structure to function, also taking life form, height, cover, shape, size, and leaf texture into consideration. The system uses letters, numbers, and drawings as symbols in such a way that the community structure is expressed through a graphic synthesis (http://www.glyndwr.ac.uk/bartlett/ecology/community.htm). For Oosting (1956), it is evidently the uniformity of the vegetation in a climatic region that is manifested in relation to the life form of the dominant species, which are specifically a product of the climate. Shimwell's system (1971) proposed the representation of vegetation characteristics through the use of formulae giving a quick and complete description of the community.

The physiognomic approach is prevalent in earth science studies (e.g. agronomics, forestry, geography) and social sciences (e.g. anthropology and archaeology) as well as among a considerable number of biologists interested in the cartographic expression of vegetation. One example of the applications is the 1:1,000,000 map produced by the United Nations Educational and Scientific and Cultural Organization (UNESCO) in 1973 (Table 2.5) with the participation of many researchers and based on the classification system proposed by Mueller-Dombois and Ellenberg (1974). From the physiognomic approach, vegetation classification systems establish links between the dominant biological forms which stand out and are chiefly the product of the relationship between the vegetation and the climate, but which invariably incorporate the distribution, abundance, and density aspects of the individuals' plants, aspects germane to the structural approach to classification systems.

The UNESCO classification is based on a hierarchy distinguished by a key of letters and numbers (first column). In the second column, come the different categories with their associated nomenclature. In the third, nomenclature is presented

Table 2.5 Some characteristics of the UNESCO 1973 physiognomic proposal (Spurr and Barnes 1982)

Class	Levels	Attribute	Level I
I, II, III, IV, V, VI, etc.	Formation class	Physiognomic	I. Closed forest
			II. Woodland
			III. Shrub
			IV. Dwarf shrub and related communities
			V. Herbaceous vegetation
			VI. Deserts and other scarcely vegetated areas
			VII. Aquatic plant formations
A, B, C, D, F, G, etc.	Formation subclass	Xeromorphy and leaf fall	
1, 2, 3, etc.	Formation group	Climate	
a, b, c, etc.	Formation	1. Leaf characteristics	
		2. Presence of conifers	
		3. Presence of spines	
		4. Succulence	
		5. Leaf width	
		6. Vegetation height	
		7. Soil characteristics	
		8. Topography	
		9. Tree crown type	
		10. Presence of bamboo	
		11. Water characteristics	
		12. Presence of palms	
(1), (2), (3), etc.	Subformation	13. Sclerophyllous floor	
(a), (b), (c), etc.	Further subdivisions	14. Presence of calcium	

for the first level (formation class), and, in the fourth, come the attributes corresponding to the different levels—for the first physiognomy and for the second (subclass of formation), taking leaf fall and the xeromorphy into account. In the third (formation groups), climate was considered, and criteria for determining formations, subformations, and other subdivisions were from 1 to 14, over and above the attributes of the preceding categories (González-Medrano 2003).

Structural

Structural classification does not generally exclude physiognomic attributes and is related to the distribution and spatial organization of the different components of the plant community. It is also largely a function of the life form of the same. The structural approach is particularly useful in areas where the flora is not well known and the structural data can be collected quickly even by inexperienced personnel (Spurr and Barnes 1982).

Lutz's phytograms (1930), for example, are polygonal figures used to characterize four factors of the physiognomic structure of species, and they allow for comparisons between the following: (1) *density*, (2) *frequency*, (3) *dominance*, and (4) *size-based classes*. The profile diagram was initially put forth by Davis and Richards (1934) and describes vegetation stratification through the use of semi-schematic —profile diagrams‖. It is applied mainly to tropical zones, though it can also be applied to temperate and boreal regions.

Larsen and Holm-Nielsen (1979) presented another structural system, which was developed for describing tropical vegetation with a view to military and counterinsurgent strategies. The descriptions are expressed through formulae obtained by assigning symbols divided into 12 groups: (1) *biological form*, (2) *predominance*, (3) *distribution*, (4) *stratification*, (5) *periodicity*, (6) *leaf characteristics*, (7) *height*, (8) *diameter*, (9) *spacing*, (10) *characteristics of the trunk and of the superficial roots*, (11) *elasticity*, and (12) *the nature of possible hindrances and dangers*.

Physiognomic–structural diagrams are, in general, stratification–cover–species–number outlines and result in a graphic representation of the horizontal strata of the vegetation. They can be used instead of profile diagrams (Table 2.6). The arrangement of a community can be schematized and described according to its physiognomy and structure, but its functional features can be another useful criterion in vegetation classification systems.

Functional

Fosberg's system is based on (1) *physiognomy (the vegetation's appearance)*, (2) *structure (the vegetation's distribution)*, and (3) *function (retention or fall of the foliage)*. This system is a diverse method widely used the world over, and it divides the vegetation successively into smaller units called **formations**. The method uses a key which first offers three alternatives leading to the first division corresponding to primary structural groups:

Table 2.6 Categories and criteria used in Dansereaugrams

Categories	Criteria
(A) Structure	1. Canopy type (open or closed)
	2. Tree spacing (uniform or irregular with distances between trees)
	3. Stratification (enumerated strata)
	4. Strata description (height and foliage)
	5. Observations on associations in strata
(B) Physiognomy	6. Height and distribution of lianas and epiphytes
	7. Trunk characteristics (e.g. buttress roots and stem succulence
	8. Observations on special life forms (e.g. arborescent ferns, cicadofitas)
	9. Evergreen or deciduous trees with foliage (percentage)
	10. If deciduous, time with foliage and time of leaf fall
	11. Leaves (simple or complex)
	12. Leaf size (uses Raunkiær's size classifications)
	13. Life forms (ground and soil strata field)
	14. Ground strata periodicity (e.g. seasonality)
	15. Seed production and method of dispersal (most important trees)
	16. Vegetative propagation (trees and bushes)
	17. Floristic composition (by stratum listed separately)
(C) Site	18. Name of the community (of dominants or characteristics)
	19. Common name (where applicable)
	20. Site location
	21. Size and type of sample (no less than half a hectare)
	22. Elevation, exposure, aspect, and slope.

Taken from Matteucci and Colma (1982)

Closed vegetation: crowns or touching stems or superimposition over 80–100 % of the cover.
Open vegetation: the plants are separated by a space at least twice their own diameter in 20–80 % of the cover.
Very open vegetation: the plants in the most complete stratum are separated by great distances and form less than 20 % of the cover.

Each primary structural group is divided into what constitutes a second tier of vegetation unit known as **classes of formation**, each of these being distinguished and receiving its name according to the dominant stratum (e.g. *dwarf shrub*). The following level is the **formation group**, in which each class of formation is divided according to function, that is to say whether the foliage is perennial or deciduous. Following on, the fourth division concerns itself with leaf texture, size, and shape in the dominant stratum, resulting in the **formations** being based on physical criteria and, sometimes, subdivided into **subformations**, which make up the fifth and last division.

This table (Richards et al., 1940) sums up the parameters considered in making a profile diagram of tropical vegetation—an example of a physiognomic–structural classification system.

As their popularity and relevance to the understanding of functional processes increased, criteria such as dispersal syndromes, establishment mechanisms, and pollination strategies, among others, began to be incorporated among the divisions, providing a very complex range of possibilities but having only limited applications, given that they require information, which, in the main, is non-existent. This functional approach is mainly suited to autophytoecological studies.

Floristic

Since the inception of Linnæus' binomial nomenclature system, information on floristic composition has been receiving greater attention. Initially, physiognomic–structural–functional classification systems gained strength using dominant species to name categories. At the beginning of the twentieth century in places where quasi-complete inventories were starting to become the norm, unlike in other less sampled regions, such as most of the tropics, there was set in motion the creation of a tool permitting not only the organization of data already gathered, but also of provision of direction and of a space for subsequent vegetation studies. The most important approach was the phytosociological, of which the principal exponent was Josias Braun-Blanquet, whose method for studying plant communities and their relationship with the terrestrial medium followed a trajectory similar to that developed for the classification and nomenclature of individual species. This system requires knowledge of the floristic composition in order to differentiate between and name the vegetation units, the basic unit being the association, a term used mainly by Humboldt (1805) and formalized by Flahault at the First International Botanical Congress held at Paris in 1900. At the sixth Congress (Amsterdam 1935), it was agreed that the term —association‖ would only be used for units defined by the possession of characteristic and differential species as per the phytosociological approach or Zürich-Montpellier school or at least as per units of the same order of importance (González-Medrano 2003).

Vegetation researchers in North America, unlike their counterparts in Europe, followed a different course when it came to the conception and use of the term —association‖. On the one hand, it can be explained through the term's being clearly identified with rigid European definitions, and, on the other, as being due to the Clementian sense used by the North Americans, which was only related to climax communities. For a time, the term fell into disuse in the English-speaking world, —community‖ being used instead. That said, although this term is not restrictive in and of itself, it was coined with an eye to its popularity. It is currently recognized that the particular meaning of —community‖ can be applied to all categories or levels of vegetation (e.g. holm–pine community, climax community—Spurr and Barnes 1982).

The phytosociological approach consists of a detailed description of the vegetation's structural morphology, its temporal dynamic, its spatial placement, its evolutionary history, and its denomination according to a well-established nomenclature for species associations (Werger 1974). The phytosociological approach, or Zürich-Montpellier school, has a solid scientific base and addresses the necessities of vegetation classification at an appropriate, efficient, and versatile level in relation to other approaches, leading to its gradual acceptance across Europe and in various regions of other continents (Africa, Asia, and in a goodly part of the Americas). In the same way, in the past decade, scientists in the English-speaking world with the partial exception of England and Australia (Schamineé et al. 1995; Jennings et al. 2002) have recognized the usefulness of the phytosociological approach in terms of the development of classification systems facilitating the characterization of vegetation cartography. In this sense, and for pragmatic reasons, it is recognized as important that phytosociology sensu stricto offers: (**I**) a minimum unit of the classification system (the association), (**II**) a sampling unit (the relevé), (**III**) a sampling method (stratifying the random), (**IV**) a single mechanism for categorical denomination (syntaxonomy), (**V**) a mechanism for territorial expression (synchorology), (**VI**) a dynamic vision of vegetation (syndynamics), (**VII**) likewise a method for describing structure (synmorphology), and (**VIII**) a temporal conception of vegetation (synchronology).

The phytosociological methodology consists of two steps: the first (analytical) involves carrying out the sampling phase, and the second (synthetic) is where the vegetation tables are derived through analogical comparison of the relevés. The sampling unit is the phytosociological relevé, and it is based on the selection of a minimum area delimited by its ecological homogeneity and uniformity in which each species' coverage is estimated as a function of the projected coverage by stratum of the whole of the individuals of the species. Moreover, in each relevé, a series of physiognomic and physiographic data taken during the sampling are linked (Werger 1974). The information obtained from these relevés is entered into a relational database matrix structure in which the information is weighted and the diagnostic species—those which determine the associations and their names are identified. This entire process is very elaborate and can be carried out by computational ordination methods and classification (Ter Braak and Prentice 1988; Velázquez 1994).

Among the data to be processed, a distinction is drawn between those of a physiognomic nature: (*a*) type of plant community, (*b*) height, (*c*) height by stratum, (*d*) dominant biotypes, (*e*) presence of bare soil and rock, the coverage by strata and by species expressed as a percentage; and those of a physiographic nature: (*a*) topography, (*b*) substratum, (*c*) elevation, (*d*) exposure, (*e*) slope, (*f*) coordinates, and (*g*) tree height. Over and above these data, the surface area of the land inventoried is estimated in the field, and all the different plant species found on the sampling site are collected for correct determination in the laboratory. Later, the series of delimited bioclimatic storeys are determined as a function of changes in temperature and precipitation (thermoclimate and ombroclimate). From these changes originate different vegetation storeys occurring from the basal zone through to the mountain tops. This phytosociological approach is used increasingly, has affected the formation of what is known today as —vegetation science‖, and is of great usefulness in addressing cartographic issues.

Landscape

Landscape is a conceptual framework derived from diverse disciplines and makes a holistic study of a territory. Within this framework, there exist different orientations, among which the most widely known are landscape ecology, geoecology, and ecogeography. Vegetation is a central element or component for all these fields and thus it has been studied from various approaches. Granados and Tapia (1990) mention that the French phytogeographical school has devised a tridimensional method, called —block diagrams‖, to represent plant communities.

The structure is a block to scale representing a mesoregion, establishing within the relief configuration, the topography, hydrological aspects, and, most importantly, the vertical and horizontal distribution of the life forms. It allows the coverage density to be observed. In the same way, succession stages in the vegetation dynamic can be considered. On the other hand, a clear representation of the distribution of vegetation types can be made for areas of lesser size in specific ways including, in many cases, urban areas (Fig. 2.2). Likewise, the soil profile, adding indicators related to drainage, salinity, stoniness et cetera, can be represented at the front of the diagram. The following are necessary features of a block diagram: aerial photography, the cartography of potential land use, and botanico-ecological knowledge of the zone in question. This information is entered into large qualitative and quantitative databases of information on plant species and the environmental factors associated with them. In general, three goals are pursued within this kind of approach: association classification plus cartographic and topological descriptions along with the identification and ordination of the factors best explaining the spatiotemporal patterns observed. Because of the associated complexity, multivariate analytical methods are used.

The development of increasingly efficient computational methods and of sophisticated statistical methods has led to an advance in the analysis and interpretation of large quantities of data through relatively few parameters—in this case, of multiple variables within ecosystems. There currently exists a great variety of methods, among them one being to characterize life forms using a multivariate approach to ordination and classification (CANOCO and TWINSPAN), which works from the basis that organisms present similar morphological adaptations, which, theoretically, at least, respond functionally to the environmental pressures to which they are subjected. These methods are an alternative to the purely taxonomic in ecological descriptions, explanations, and preconditions of plant communities (Hill 1979; Ter Braak 1986; Kent and Coker 1992; Velázquez and Cleef 1993; Dallas 2000).

Fig. 2.2 Degree of anthropogenic influence across the landscape results in complex intermingling mosaic vegetation formations. These range from predominantly native plant species aggregates (where human actions are mostly subordinated to natural conditions) to predominantly man-made plant species aggregates (where human actions are the system's driving force) (Taken from Zonneveld I. 1979. Rural Ecology Course lecture notes, ITC-The Netherlands, unpublished notes)

According to Matteucci and Colma (1982), there exist two grand approaches to vegetation ordination. The first type deals with the use of mathematical tools for the reduction of the —dimensionality of the vegetational space‖, referred to as —**taxometric ordination**‖ and consisting of the ordination of the set of individuals recognized in the sampling exercises. In the second type, referred to as —**ecological ordination**‖, the attempt is made to find relationships between variations in the vegetation (structural descriptors of the community) and the environmental gradients. In these approaches, the authors recommend —sacrificing mathematical exactitude in favour of ecologically interpretable results‖.

Another plant community classification technique is that put forth in the theory of fuzzy sets, which allows for the inclusion of certain attributes not easily incorporated into conventional approaches to classification, namely because plant communities are not separated by precise limits, but, on the contrary, there is a blurring of the lines, rendering overlap and internal heterogeneity characteristic of ecological communities. The concepts of ecosystem as well as those of community, association, and trophic level, et cetera, are all examples of fuzzy sets. The theory of fuzzy sets is useful for description and analytical purposes in these circumstances and serves as an alternative conceptual base for the most satisfactory representation of ecological communities (Equihua 1991; Pérez-Vega et al. 2010).

Predominant Approaches in Mexico

The Physiognomic–Structural–Floristic Approach

In the Mexican context, the first attempts at vegetation classification were based on physiognomic features (Sanders 1921; Shelford 1926 and Smith and Johnston 1945, cited in Rzedowski 1978). Even so, said works also tend to use bioclimatic or geomorphological criteria and lay out features of the floristic composition in their characterization. Leopold (1950) outlines a table distinguishing between 12 types of vegetation and indicates their distribution on a map originally published on a scale of 1:20,000,000. Miranda and Hernández-X (1963) recognize 32 types of vegetation. The work lacks a map but includes data on geographic distribution, principal relationships with environmental factors, photographs, and a succinct description of the floristic composition of each type of vegetation included. Flores-Mata et al. (1971) prepared an excellent cartographic reference in which 25 categories were recognized including as many vegetation types as land uses. That said, the document does not explain many of the criteria selected in denominating the various categories.

Rzedowski (1978) relates his proposal with other classification systems, amply justifies the categories created, provides an outline distribution map, and describes in great detail the most outstanding features of the climate and of the floristic composition differential of his proposed categories in differing regions and entities of the country as well as the general difficulty arising from the lack of information and/or exploration or deterioration accelerated by human activity. This book also includes vegetation research material and classification diagrams for the country as a whole and by individual region within his proposed categories and dedicates ample discourse to climate, physiography, and possible floristic relationships, making this indubitably the most complete contribution on Mexican vegetation.

The soil and vegetation classification system designed and used by the Instituto Nacional de Estadísticas y Geografía (INEGI 1999) separates the physiognomic and structural features of vegetation types from land use. Later, there appears the proposed —communities' used in the Inventario Nacional Forestal (INF) of 2000 (Palacios-Prieto et al. 2000), in which are incorporated criteria related both to the classification of coverage types and to land use as different physiognomic–structural and compositional features of plant community typification.

The 2003 González-Medrano proposal for unified plant community classification criteria in Mexico also stands out, given that it makes clear that the criteria regulating the proposal up to the formation level are bioclimatic–

physiognomic/structural and that they correspond approximately with numerous categories of vegetation types used in aforementioned systems insofar as the criterion of floristic composition is incorporated in the proposal at the levels of association and consociation. Unfortunately, the work does not go as far as characterizing any particular area of Mexico at these levels.

The prevalent approach in Mexico is physiognomic–structural–floristic. Even so, the databases in general are descriptive in nature, and thus the categories used to define vegetation types are very different. It would be impossible to include a highly detailed description of the multiple proposals drawn up for the various regions of the country. In general, there are five to a greater or lesser degree prevalent and fundamental to the creation of adaptations of higly specific proposals where they are needed to resolve a particular situation. These proposals are described in Table 2.7, in which the attempt has been made to find the level and nomenclature corresponding to that of the various authors. It should be mentioned that this table was originally drawn up by Dr. J. Rzedowski, and, in this present work, only previously non-existent information has been added—for example, the term —Xerophile shrubland‖ has its equivalent in various of Miranda and Hernández-X's 1963 plant-classification units, indicating that they are not the same category.

The preceding table raises the general notion that each and every one of the different units described can be placed in a hierarchy of distinct levels. That said, different criteria are used from system to system and sometimes even within the same system, making it difficult to understand the hierarchy and structure derived from the proposals. A more precise example of levels and associated nomenclature originating from some case studies on particular regions is shown in Table 2.8.

It should also be mentioned that the contributions made on this subject by Mexican governmental institutions carrying out research on vegetation are fundamental source material, given the importance of the research derived from its national scope. The same can be said of research carried out by environmental protection organizations. Proposals originating with institutions seek, in the main, to create cartographic expressions of vegetation in contrast to earlier studies, the aim of which had been to study vegetation in its chorological expression. Outstanding among this research are the efforts of various bodies: the Secretaría de Programación y Presupuesto (SPP), which publishes its charts on land use and vegetation. It has about 80 terms its uses to describe different vegetation communities, many of which are taken from the works of various researchers, but the majority of them are adequate for the secretariat's own purposes. The INEGI in its various publications, such as its 1:250,000-scale land use and vegetation charts, uses terminology consistent with Miranda and Hernández X's 1963 proposal, all the while adapting it for the institute's own purposes. The INEGI uses over 60 categories. Additionally, for the regional determination of Rangeland Coefficients, (COTECOCA), the Comisión Técnica Consultiva uses approximately 70 categories to describe units of vegetation. In its turn, the Instituto de Geografía at UNAM, in its bulletin —*Investigaciones Geográfica‖*, 43 (Palacios-Prieto et al. 2000), recorded the terms **—formation‖**, **—types of vegetation‖**, **—communities‖**, and **—subcommunities‖**, terms which would later be used by the National Forest

Table 2.7 Approximate correspondence between nomenclatural terms in different vegetation classification systems in Mexico. Rzedowski (1978) was taken as the starting point

Rzedowski (1978)	Miranda y Hernández X. (1963)	Flores-Mata et al. (1971)	Palacios-Prieto et al. (2000)	González-Medrano (2003)
Evergreen tropical forest	High evergreen jungle, high or mid-height subevergreen jungle	High evergreen jungle	Evergreen jungle and subevergreen	Humid tropical forest (in part)
Subdeciduous tropical forest	High or mid-height subdeciduous jungle	Mid-height subevergreen jungle (in part)	Deciduous jungle and subdeciduous (in part)	Humid tropical forest (in part)
Deciduous tropical forest	Low-height deciduous jungle	Low-height deciduous jungle (in part)	Low deciduous jungle (in part)	Dry tropical forest (in part) Tropical mountain forest?
Spiny forest	Low subevergreen jungle (in part), Low spiny evergreen jungle, Low spiny deciduous jungle?	Low deciduous jungle (in part), low subevergreen jungle, Mezquital (in part)	Mezquital (in part)	Dry tropical forest
Grassland	Grassland, zacatonal, high paramos vegetation	Meadowland, zacatonal, savannah	Meadowland	Meadowland
Xerophile shrubland	Spiny shrub with lateral spines, cardonales, Neobuxbaumia tetetzo, etc., izotal, nopaleras, spiny shrub with terminal spines, spineless parvifolia shrub, magueyales, agave, lechugilla, Hechtia glomerata, etc., chaparral, arid sandy desert vegetation	Mezquital (in part), chaparral, submountainous shrub, fleshy stem shrub, desert shrubland, rosetophilous shrubs, micropylar desert shrubland	Mezquital (in part) Xerophile shrubland	Xerophile shrubland

(continued)

Table 2.7 (continued)

Rzedowski (1978)	Miranda y Hernández X. (1963)	Flores-Mata et al. (1971)	Palacios-Prieto et al. (2000)	González-Medrano (2003)
Quercus forest	Holms	Oak forest	Hardwood forest	Temperate forests (part) Non-hydrophilic forest (in part)
Coniferous forest	Pines, fir forest	Pine forest, fir forest	Coniferous forest, coniferous–hardwood forest	Temperate forests (part) Non-hydrophilic forest (in part)
Mountain cloud forest	Mid or low evergreen jungle, deciduous forest (in part)	Deciduous forest	Mountain cloud forest	Tropical mountain forest?
Aquatic and subaquatic vegetation	Mangrove swamp, Thalia geniculata, tulares, reed beds, etc., deciduous forest (in part)	Mangrove, Thalia geniculata	Hydrophilic vegetation	Hydrophilic forests Freshwater hydrophilic vegetation

Inventory. As can be seen in the aforementioned examples, the physiognomic, structural, and floristic (not necessarily in that order) characteristics are those most prevalent.

Phytosociological Approach

Rigorous research from the phytosociological approach in Mexico is scarce. In general, the regions of the Baja California Peninsula (Delgadillo et al. 1992; Alcaraz et Delgadillo 2010) and the centre of the Trans-Mexican Volcanic Belt (Velázquez and Cleef 1993) stand out. There also exist isolated studies on coastal dunes (Morena-Casasola et al. 1982; Castillo et al. 1991), humid tropical regions (Guevara et al. 1994; Sánchez-Sánchez and Islebe 2002), and dry tropical regions (Pérez-Vega 2008).

In the phytosociological analysis of the salt marshes and mangrove swamps of Baja California, Delgadillo et al. (1992) found a total of 16 **associations**, which were later covered by Peinado et al. (1994b). Another phytosociological study was carried out by Peinado et al. (1994a) in the Sierra de San Pedro Mártir in Baja California. It describes five vegetation storeys and provides their climatic characteristics, denominated as follows: Infra-Mediterranean, Thermo-Mediterranean,

Table 2.8 Some contributions to knowledge on the classification of vegetation from the prevalent school in Mexico

Author(s)	Region or location of study	Levels and nomenclature
Zamudio et al. (1992)	Querétaro State	Level I: Vegetation types
		Level II: Vegetation types
		Level III: Others vegetation types
Ramos and Novelo (1993)	Yuriria Lagoon, Guanajuato	Level I. Associations
González-Elizondo et al. (1993)	"La Michilía" Biosphere Reserve, Durango	Level I. Temperate forest
		Level II. Vegetation types
		Level III. Units of vegetation
		Level IV. Associations
Madrigal (1994)	Eastern Michoacán	Level I. Formation
		Level II. Subformation
		Level III. Associations
		Level IV. Series
		Level V. Facies
Labat (1995)	North-eastern Michoacán	Level I. Formation
		Level II. Groups
		Level III. Communities
Pennigton and Sarukhán (1998)	Hot humid zone of Mexico	Level I. Vegetation types
Rzedowski and Rzedowski (2001)	Valle de México	Level I. Vegetation types
		Level II. Others Vegetation types
		Level III. Associations
		Level IV. Communities
Briones and Villarreal (2001)	Vegetation and flora of on ecotone between the province of "*Altiplano*" and "*Planicie Costera*" in the north-east of Mexico	Level I. Vegetation types

Meso-Mediterranean, Supra-Mediterranean, and Mesotropical. In each of these levels, units of vegetation are identified using the term **association** as per phytosociological methods. Later, Rivas-Martínez (1997) and Rivas-Martínez et al. (1998, 1999b) determined a series of bioclimatic storeys delimited by temperature changes and precipitation (thermoclimate and ombroclimate) for North America, and, in the same way, presented a phytosociological synthesis of forest vegetation in the north of Mexico which includes six **classes** and eight **orders**, 19 **alliances,** and 64 **associations**. Guevara et al. (1994) examined the vegetation and flora of paddocks in the Sierra de los Tuxtlas, obtaining, as a result, floristic groupings, which they called **floristic sets**.

Few studies taking this approach maintain the regulation available regarding the levels of aggregation and nomenclature. Outstanding among these studies are those of Almeida-Leñero et al. (1994) and Almeida-Leñero (1997), in which they report on various levels of **order**, **alliance**, **community**, **association,** and **subassociation** Three further examples are those of Giménez et al. (1997, 2003—Monarch Butterfly Region) and González et al. (2002—Sierra de Catorce).

The Challenge of Producing an Integrative and Inclusive Proposal

The preceding is but an outline of what exists in Mexico in terms of vegetation studies emphasizing the classification and nomenclature aspects of the object of study. The number of scientific contributions would be overwhelming if all the taxonomic, floristic, and especially ecological researches were totalled. An example of this can be seen in the publications of various universities, both national and state, which have accumulated hundreds, if not thousands, as in the case of UNAM, of theses and articles. At the same time, enormous databases such the ones available for consultation at CONABIO (http://www.conabio.gob.mx) and INECC (http://www.inecc.gob.mx) are examples of the great effort being made to systemize existing information on vegetation and its flora. Another piece of evidence is the number of works presented in various botanical fora in the form of both national and international congresses. Faced with this situation, it is crucial to have a system which does not disown experience already accrued and which systemizes existing knowledge, but which is also congruent and consistent with the levels, criteria, and nomenclature of the various categories of vegetation. The final aspect is of capital importance, given that nomenclature becomes an index of the degree of beta diversity which is established based on the number of native vegetation types found in the region of study (e.g. the province, municipality, or watershed). The inconsistent use of terms results in vocabulary incompatibility between regions as illustrated below:

1. The use of toponymic terms—for example, the —Matorral Tamaulipeco‖ to describe the shrubland found in the Tamaulipas State, the term implying the shrubland found there is somehow unique to the region (e.g. Dinerstein et al. 1995).
2. The use of the dominant tree genus in preference to the phenology. The case of —Holm forest‖ with at least three connotations—deciduous, perennial, and subperennial, which implies a reduced diversity through the sole use of the genus (Rzedowski 1978).
3. The use of the mesoclimatic condition, the most obvious example being that of —Mesophyll forest‖, which denomination refers to the leaf type, when, in reality, it is making reference to the type of mesoclimate. Other examples are —Xerophile shrubland‖ and —Arid desert vegetation‖ (Rzekowski 1978; Miranda and Hernández X 1963).
4. The use of geomorphological or physiographic terms—again bringing up the terms —Mountain mesophyll forest‖, —Alpine prairie‖, —Coastal dunes‖, and —Arid sandy desert vegetation‖ (Miranda 1947; INEGI 1999).
5. The use of local names or words such as —Tetechera‖ (named after the dominant species: *Neobuxbaumia tetetzo*), —Nopalera‖, —Popal‖, —Tular‖, —Izotal‖, and —Carrizal‖, among others (Miranda and Hernández X 1963).

To be complementary, the ideal system would need to be compatible with the systems developed for other parts of the world, especially North America, inclusive of Central America, in order to meet macroregional demands as an effect of global change. Additionally, the proposal for the ideal system derived from this review would need to address the cartographic expression of the categories, that is to say the creation at the same time of legends on various scales, which would allow for consistency between the classification system and its respective proposed legend at any given scale. This concept will be developed in the following chapters and represents the core of this work.

References

Alcaraz F, Delgadillo J (2010) Matorrales suculentos y desérticos del desierto central de Baja California afectados por la maresìa. Braun-Blanquetia 46: 85–100.

Almeida L, Cleef AM, Herrera A, Velázquez A, Luna I (1994) El zacatonal alpino del Volcán Popocatépetl, México y su posición en las montañas tropicales de América. Phytocoenologia 22(3): 391- 436.

Almeida-Leñero (1997) Vegetación, fitogeografía and paleoecología del zacatonal alpino and bosques montanos de la región central de México. Tesis doctoral, Universidad de Amsterdam.

Archibold OW (1995) Ecology of World Vegetation. Chapman & Hall, London.

Bojórquez-Tapia LA, Azuara I, Escurra E (1995) Identifying conservation priorities in Mexico through geographic information systems and modeling. Ecol. Applic. 1: 215–231.

Briones O, Villarreal JA (2001) Vegetación and flora de un ecotono entre las provincias del Altiplano and de la Planicie Costera del Noreste de México. Acta Botánica Mexicana. Instituto de Ecología A.C. Pátzcuaro, Michoacán México, 55: 39–67.

Cain SA (1932) Concerning certain phytosociological concepts. Ecological Monographs 2: 475–508.
Cain SA, Castro GM (1959) Manual of vegetation analysis. Harper & Brother, New York.
Carabias J, Landa R (2006) Agua, medio ambiente and sociedad: hacia una política integral de los recursos hídricos. El Colegio de México-UNAM-Fundación Gonzalo Río Arronte, México.
Castillo S, Pompa J. Moreno-Casasola P (1991) Coastal sand dune vegetation of Tabasco and Campeche, Mexico. Journal of Vegetation Science 2: 73–88.
Dallas EJ (2000) Métodos Multivariados Aplicados al Análisis de Datos. Internacional Thomson, México.
Dansereau P (1951) Description and recording of vegetation upon a structural basis. Ecology 32: 172–229.
Davis TAW., Richards PW (1934) The vegetation of Moraballi Creek, British Guiana; and Ecological Study of a Limited Area of Tropical Rain Forest. I and II, J. Ecol. 21: 350–384 (1933) and J. Ecol. 22: 106–155.
Delgadillo J. Peinado M, De La Cruz M, Martínez-Parras J, Alcaraz F., De La Torre A (1992) Análisis fitosociológico de los saladares and manglares de Baja California, México. Acta Botánica Mexicana. Instituto de Ecología, A. C. Pátzcuaro, Michoacán. México, 19: 1–35.
Dinerstein E, Olson D. Graham D, Webster A, Primm S, Bookbinder M, Ledec C (1995) Una evaluación del estado de las eco-regiones terrestres de América Latina and el Caribe. Banco Mundial, WWF. Washington D.C.
Equihua M (1991) Fuzzy clustering of ecological data. Journal of Ecology, 78: 519-534.
FGDC (Federal Geographic Data Committee) (2005) Nacional Vegetation Classiffication Standard Hierarchy Revisions Working Group. Federal Geographic Data Committee, Vegetation Subcommittee. OverView July 2005. In: http://biology.usgs.gov/fgdc.veg.
FGDC (Federal Geographic Data Committee) (2008) National Vegetation Classification Standard, Version 2. Vegetation Subcommittee. Federal Geographic Data Committee.
Flores-Mata G, Jiménez L, Madrigal S, Moncayo R., Takaki T (1971) Tipos de Vegetación de la República Mexicana. Secretaría de Recursos Hidraúlicos. México, D.F.
Fosberg FR (1961) A classification of vegetation for general purposes. Trop. Ecol. 2:1–28 (Republished in modified form in Peterken GF 1967. Guide to the Check Sheet for IBP Areas. IBP Handbook 4. Blackwell Scientific. Oxford: 73–120).
García E (2004) Modificaciones al Sistema de Clasificación Climática de Köppen. Instituto de Geografía. Universidad Nacional Autónoma de México. Serie Libros, 6. México, D.F.
Giménez J, Escamilla M, Velázquez A (1997) Fitosociología and sucesión en el volcán Paricutín, Michoacán, México. Caldasia 19 (3): 487–505.
Giménez JM, Ramírez IM, Pinto M (2003) Las comunidades vegetales de la Sierra de Angangueo: clasificación, composición and distribución. Lazaroa 24: 87–111.
González-Medrano F (2003). Las comunidades vegetales de México.Propuesta para la unificación de la clasificación y nomenclatura de la vegetación de México. Instituto Nacional de Ecología, México, DF.
González-Elizondo S, González-Elizondo M, A Cortés O (1993) Vegetación de la Reserva de la Biosfera - La Michilía, Durango, México. Acta Botánica Méxicana. Instituto de Ecología, A.C. Pátzcuaro, Michoacán. México. 22: 1–104.
Good R (1953) The geography of flowering plants. II ed., Longmans, Green & Co., London.
Granados SD, R Tapia V (1990) Comunidades vegetales. Universidad Autónoma de Chapingo. México, D.F.
Guevara S, Meave J, Moreno-Casasola P, Laborde JJ (1994) Vegetación and flora de los potreros en la sierra de los Tuxtlas, México. Acta Botánica Méxicana. Instituto de Ecología, A.C. Pátzcuaro, Michoacán. México, 28: 1–27.
Hill MO (1979) TWINSPAN - A FORTRAN program for arranging multivariate data in an ordered two-way table by classification of individuals and attributes. Cornell. University Press, Ithaca N.Y.
Holdridge LR (1967) Life zone ecology, San José, Costa Rica.

Holdridge LR, Grenke CW, Hatheawt HW, Liang T, Tosi AJ (1971) Forest environments in tropical life zones a pilot study. Pergamon Press., New York.

Huber O, Riina R (2003) Glosario Fitoecológico de las Américas. Vol. 2. México, América Central e Islas del Caribe: países hispanoparlantes. Ediciones UNESCO. Corolab. Humboldt, Paris.

Humboldt A (1805) Essai sur la geographic des plantes. In Voyage, Pt. 5, Physique generale et geologie, Paris.

INEGI (Insitituto Nacional de Estadística Geografía e Informática). (1999) Información Botánica en la cartografía temática del INEGI. Guía Normativo-Metodológica. Dirección General de Geografía INEGI. México.

Jennings M, Loucks O, Glenn-Lewin D, Peet R, Faber-Langendoen D, Grossman D, Damman A, Barbour M, Pfister R, Walker M, Talbot S, Walker J, Hartshorn G, Waggoner G, Abrams M, Hill A, Roberts D, Tart D. (2002) Standards for associations and alliances of the U.S. national vegetation classification. Version 1.0. Vegetation Classification Panel, Ecological Society of America. Washington, D.C. USA.

Kent M, Coker P (1992) Vegetation description and Analysis. A Practical Approach. Belhaven Press, London.

Küchler AW (1949) A physiognomic classification of vegetation. Ann. Ass. Am. Geogr. 39: 201–210.

Labat N (1995) Végétation du Nordouest du Michoacán, Mexique. Instituto de Ecología A.C.

Larsen K, Holm-Nielsen B (eds) (1979) Tropical botany. Academic Press, London.

Leopold AS (1950) Vegetation zones of Mexico. Ecology. 31: 507–518 (1735–1770).

Lutz HJ (1930) The vegetation of Heart's Content, a virgen forest in northwestern Pennsylvania. Ecology II: 1–29.

Madrigal X (1994) Características generales de la región forestal oriental del estado de Michoacán, México, UMSNH Morelia, México.

Matteucci SD, Colma A (1982) Metodología para el estudio de la vegetación. Serie Biología OEA. Monografía 22. Washington.

Miranda F (1947) Estudios sobre la vegetación de México. Rasgos de la vegetación en la Cuenca del Río de las Balsas. Rev. Soc. Mex. Hist. Nat. 8: 95–114.

Miranda F, Hernández XE (1963) Los tipos de vegetación de México and su clasificación. Bol. Soc. Bot. Méx. 28: 29–179.

Moreno-Casasola PE, van der Maarel E, Castillo S, Huesca ML, Pisanty I (1982) Ecología de la vegetación de dunas costeras: estructura and composición en el Morro de la Mancha, Ver. I. Biótica 7: 491–596.

Mueller-Dombois D. Ellenberg H (1974) Aims and Methods of Vegetation Ecology. John Wiley & Sons, Manchester.

Olson D, Dinerstein E, Wikramanayake D, Burgess D, Powell N, Underwood C, D'Amico A, Itoua I, Strand H, Morrison C, Loucks J, Allnutt F, Ricketts H, Kura Y, Lamoreux F, Wettengel W, Hedao P, Kassem R (2001) Terrestrial Ecoregions of the World: A New Map of Life on Earth. BioScience, 51(11): 933-938.

Oosting HJ (1956) The study of plant communities. W. H. Freeman and Co., San Francisco.

Palacios-Prieto JL, Bocco G, Velázquez A, Mas JF, Takaki-Takaki F, Victoria A, Luna-González L, Gómez-Rodríguez G, López-García M, Palma M, Trejo-Vázquez I, A Peralta H, Prado-Molina J, Rodríguez-Aguilar A, Mayorga-Saucedo R, F González M (2000) La condición actual de los recursos forestales en México: resultados del Inventario Forestal Nacional 2000. Investigaciones Geográficas, Boletín del Instituto de Geografía. UNAM. 43: 183–203.

Peinado M, Bartolomé C, Delgadillo J, Aguado I (1994a) Pisos de vegetación de la Sierra de San Pedro Mártir, Baja California, México. Acta Botánica Mexicana. Instituto de Ecología A.C. Centro Regional del Bajío. Pátzcuaro, Michoacán. México, 29: 1–30.

Peinado M, F. Alcaraz JL, Aguirre & Alvárez J (1994b) Vegetation formations and associations of the zonobiomes along the North American Pacific Coast: From Northern California to Alaska. Plant Ecol. 129: 29–47.

Pérez-Vega A (2008) La contribución de los atributos del relieve para explicar la distribución espacial de la vegetación: estudio de caso en una zona de montaña de alta diversidad. Tesis de Doctor en Geografía. Universidad Nacional Autónoma de México. Morelia, Mich.

Pérez-Vega A, Velázquez A, Giménez de Azcárcate J (2010) Aproximación geobotánica de bosque tropical seco caducifolio en la microcuenca del Río San José de Chila. Michoacán, México. Braun-Blanquetia 46: 351–359.

Ramos L, Novelo A (1993) Vegetación y flora acuática de la laguna de Yuriria, Guanajuato, México. Acta Bot. Mex, 25: 61–79.

Raunkaier C (1934) The life forms of plants and statistical plant geography. Clarendon, Oxford.

Richards, P.W., A.G. Tansley and A. S. Watt. 1940. The Recording of Structure, Life Form and Flora of Tropical Forest Communities as a Basis for Their Classification. Journal of Ecology 28(1):224-239.

Riley D, Young A (1966) World Vegetation. Cambridge University Press. London.

Rivas-Martìnez S (1997) Syntaxonomical synopsis of the potential natural plant communities of North America, I. Itinera Geobot. 10: 5–148.

Rivas-Martìnez S, Giménez de Azcárate J, Penas-Merino A, Palma-Muñoz M (1998) Esbozo bioclimático de México. In: Libro de Resúmenes VII Congreso Latinoamericano de Botánica. XIV Congreso Mexicano de Botánica. Sociedad Botánica de México. México, D.F.: 201–202.

Rivas-Martímez S, Sánchez-Mata D, Costa M (1999) North America boreal and western temperate forest vegetation (Syntaxonomical synopsis of the potential natural plant communities of North America II). Itinera Geobot. 12: 5-316.

Rzedowski J (1978) Vegetación de México. Limusa. México, D.F.

Rzedowski C, Rzedowski J (2001) Flora fanerogámica del Valle de México. Instituto de Ecología y Comisión Nacional para el Conocimiento y Uso de la Biodiversidad. Pátzcuaro, Michoacán, México.

Sánchez-Sánchez O, Islebe GA (2002) Tropical forest communities in southeastern Mexico. Plant Ecology Kluwer Academic Publishers, Plant Ecology 158: 183–200.

Sanders EM (1921) The natural regions of México. Geog. Rev. 11: 212–226.

Schaminée JHJ, Weeda EJ, Westhoff V (1995) De vegetatie van Nederland. Deel 2. Plantengemeenschappen van wateren, moerassen en natte heiden. Opulus Press, Uppsala-Leiden.

Shelford E (1926) Terms and concepts in animal ecology. Ecology, 7(3): 389.

Shimwell DW (1971) Description and classification of vegetation. Sidgewick and Jackson, London.

Smith AC, Johnston IM (1945) A phytogeografic sketch of Latin America. Plants and plant science in Latin America, 11–18.

Spurr SH, Barnes BV (1982) Ecología forestal. A.G.T. Editor, S. A. México, D.F.

Takhtajan A (1986) Floristic regions of the World. University of California Press, Berkeley, Los Angeless, London.

Ter Braak CJF (1986) Canonical Correspondence Analysis: a new eigenvector technique for multivariate direct gradient analysis. Ecology 67: 1167–1179.

Ter Braak CJF, Prentice IC (1988) Theory of gradient analysis. Adv. Ecol. Res. 18: 271–317.

Velázquez A, Cleef AM (1993) The plant communities of the volcanoes Tláloc and Pelado, México. Phytocoenología 22: 145–192.

Velázquez A (1994) Multivariate analysis of the vegetation of the volcanoes Tláloc and Pelado, Mexico. Journal Vegetation Science 5: 263–270.

Velázquez A, Romero FJ (eds) (1999) Biodiversidad de la región de montaña del sur de la Cuenca de México: bases para el ordenamiento ecológico. UAM-X-SEMARNAP. ISBN: 754 24 2867 8.

Velázquez A, Mas JF, Fiaz-Gaellgos JR, Mayorga-Saucedo R, Alcántara PC, Catsro R, Fernández T, Bocco G, Palacio JL (2002) Patrones y tasas de cambio del uso del suelo en México. Gaceta Ecològica 62: 21–37.

Velázquez A, Durán E, Mas JF, Bray D, Bocco G (2006) Situación actual and prospectiva del cambio de la cubierta vegetal and usos de suelo en México. En México ante los desafíos del desarrollo del Milenio. Consejo Nacional de Población. 391–416.

Velázquez A, Mas JF, Bocco G, Palacio-Prieto JL (2010a) Mapping Land Cover Changes in Mexico, 1976-2000 and applications for guiding environmental management policy. Singapore Journal of Tropical Geography 31: 152–162.
Werger A (1974) On concepts and techniques applied in the Zurich-Montpellier method of vegetation survey. Bothalia, 11: 309-323.
Whittaker R (1962) Classification of natural communities. Botanical Review, 28: 1–239.
Zamudio R, Rzedowski J, Carranza E, Calderón G (1992) La vegetación en el estado de Querétaro. Instituto de Ecología, A.C.- CONCYTEQ.

Chapter 3
Proposal for a Standardized Hierarchical System for the Classification of Vegetation in Mexico

Introduction

Classification *sensu lato* pursues the recognition of categories, the hierarchical arrangement expressing the (dis)similarity between them, and the unique and distinct denominations between levels of division (Anonymous 2003). These principles have been applied across various disciplines related to biology, such as taxonomy, systematics, and phylogeny, which have taken on these tasks from a precise definition of the level of specific organization. At this level, the central element is constituted by the species, for which there exist criteria characterizing them with increasing rigour and with wide universal acceptance. In the traditional sense, biology recognizes organizational levels. In this way, a set of species is aggregated into *Populations*; these constitute *Communities*, which, in their turn, form *Habitats*, and these intermingle to define *Ecosystems*, which, at a higher level, join to make up *Biomas* (Levin 2009). The criteria defining a community and what separates it from others as well as the differences between habitats, ecosystems, and biomes have been widely discussed. On the one hand, their value as natural units subject to the laws of evolution is examined, while on the other, the criteria defining them generally conform to criteria of regional value and thus do not receive universal acceptance. In this way, there is no global method designating a classification mechanism which precisely distinguishes between units, categories, a hierarchical order and a unique denomination, and which is of universal value to the different levels of biological organization above the level of populations.

Vegetation, by definition, does not escape this reality, as it is based on sets of species, communities, habitats, ecosystems, and biomes. The particular conditions in each region make it necessary to include specific modifications, as illustrated in previous chapters. In such a situation, the great challenge before making a new proposal on the subject of vegetation classification in Mexico consists of four fundamental facets: (**I**) the recognition of previous efforts, thus building upon their strengths, (**II**) the reorganization of available information in order to illustrate

A. Velázquez et al., *Standardized Hierarchical Vegetation Classification*,
Geobotany Studies, DOI 10.1007/978-3-319-41222-1_3

explicitly the mechanics of affinity between the new proposal and previous research, (**III**) highlighting the inclusive character, and (**IV**) taking into account the compatibility between the new proposal and current developments across the world, particularly in North America.

In this chapter, the indisputability is put forth of the need to create a proposal for a vegetation classification system for Mexico which satisfies the demands not only from the academic side but also from the practical side. It should be clarified that the particularities of biological diversity in Mexico presuppose that there will be many exceptions. This fact, however, cannot be used as an argument for creating a system dependent on the site and author—that is to say that, a system which ignores previous efforts and is predestined to generate incomparable results, a system of great value locally, but which is of little use at the regional or macroregional level. Within this framework described above, the central aim of the new proposal seeks to identify, characterize, and understand the composition and function of vegetation and its distinct steps of organization. Such a task gives rise to three relevant points to consider:

The first deals with the transitory or dynamic character of vegetation, given that it is undergoing degrees of transformation without precedent in the evolutionary history of life. For this reason, its study requires two different forms of looking at its dynamics. One, of purely natural origin and thus of profound ecological interest, includes the processes of succession and resilience. The other, of purely anthropogenic origin, refers to the degree to which man transforms it (secondary vegetation). It is recognized that to differentiate between the origins of vegetation disturbance and, consequently, to know whether the disturbance refers to vegetation in a succession stadium *sensu stricto* or simply to secondary vegetation is a complex task. At the local level (working with very fine scales greater than 1:25,000), the origin of the disturbance is discernible, given that the species sets will be indicative of this process. At the regional level, however, to discern between succession stadia and patches of secondary vegetation is not trivial and is sometimes unnecessary. Other parameters such as degrees of canopy fragmentation and the proportion of cultural as against natural coverage, among others, turn out to be better indicators of the type of dynamic affecting the vegetation. This indicates a radical change given that the concept of vegetation is substituted by that of plant cover. The study of vegetation dynamics *sensu stricto* is the concern of biology in its various branches, such as floristics, botany, and ecology. The analysis of the dynamics of plant cover *sensu stricto* is the domain of geography in its various sources and tools, such as remote sensing and geographical information systems, land use, and thematic chorology. The cartographic expression of vegetation, however, requires the work to be carried out in line with both concepts. This chapter limits itself to the provision of a classification framework which helps reconcile the botanical and geographical approaches.

The second point refers to the constant presence of exceptions, a product of vegetation's evolutionary history in Mexico. It is a response to the physical environment's complex history (Ferrusquía 1998) as expressed through capricious physiognomic, phenologic, and floristic characteristics (Rzedowski 1978). One part of the expression is intertwined and correlates with the dominant mesoclimate of the region. Another part responds complementarily to the geological, lithological, geomorphologic, edaphic, and hydric history of each location, which, on occasion, can be at odds with the mesoclimate. The most enigmatic example of this is the—Mesophyll forest‖ (Luna et al. 2001). Within this context, in certain regions of Mexico, the supposition that each combination of factors in the physical–climatic environment corresponds with a unique and distinct vegetation type is not valid. In these —whismsical regions‖, exceptions (azonal vegetation) are the rule, and, to a large extent, the area in which they are established is always significantly less than the space occupied by the zonal vegetation.

The third point to highlight is the historical or biogeographical aspect. As is well known, Mexico is the centre of origin, speciation, and dispersal for a large number of taxa and thus a place where endemism abounds (Rzedowski 1991a, b; Wendt 1998; Challenger 1998). This is expressed through a diversity of vegetation types where plant species of remote, relictual, and autochthonous origin intermingle, further complicating the task of classification. Much has been written on this subject, and three examples were selected to illustrate it. The first concerns the pine family (Pinaceae), which, along with the chestnut and oak or holms (Fagaceae), shares Eurasian origins. Their centre of diversification, however, is limited to Mexico, given that 160 of 500 *Quercus* species and 49 of 120 *Pinus* species worldwide are found there. A third example includes the paper barktree (*Bursera* spp), the aromatic wood of which copal is used in the manufacture of incense. There are about 100 species spread across the world, with 37 endemic to Mexico (Rzedowski 2005).

Given these three situations, it should be clarified that this work seeks primarily to deal with the classification of predominantly primary vegetation types or those with some degree of disturbance, which has allowed for insertion into a natural system under conditions either zonal or dominant in the landscape. The historical (biogeographical) aspect is inevitable and is expressed through the complex argument of vegetation types, which, in general, can be related to a logical system of mesoregional validity (North America inclusive of Central America).

In this way, the quest is for a proposal allowing the systematization, generation, and organization of information on vegetation from irrefutable criteria, which will be of eventual assistance in generating synthetic databases sufficiently solid in their concepts but, at the same time, flexible enough to be able to be adequately adapted to local and regional situations, hence taking into account the enormous variation in plant communities present in Mexico. The final goal was to construct a —system‖ of practical use, —hierarchical‖ by its very nature and —standardized‖ through its allowing for the definition of unique categories of nomenclature for each type of vegetation. The system proposed is the **Sistema de Clasificación de la Vegetación de México** (Vegetation Classification System for Mexico—SECLAVEMEX).

Conceptual Framework

Four classification systems for vegetation in Mexico will be selected based on the greatest number of citations or on their breadth in terms of covering all parts of Mexico. From these will be identified the criteria, levels, and nomenclature to consider in a new proposal. A synthesis of this revision is described in Appendix A.3.

This analysis permits the identification of common criteria and hierarchies between the systems. In this way, the growth form (or dominant life form) becomes the first and prevailing criterion denoting a physiognomic characteristic. The second criterion corresponds to the fall or retention of foliage, permitting a phenological separation for a large number of categories, occasionally as an indicator of the dominant mesoclimate. Other criteria do not appear to be shared either in relevance or in hierarchy. Following on from this analysis, there arises the necessity of organizing criteria, defining and proposing levels and coining names for the different hierarchical levels with the aim of satisfying the requirements of new a classification system.

Criteria

Nine characteristics were chosen which correspond with botanical characteristics commonly used to discern different types of vegetation. Criteria of climatic context were also chosen for the close relationship between the vegetation distribution patterns and the climates. Finally, also included is the criterion of the condition expressed in the degree of disturbance or naturalness. It should be clarified that any change, addition, or exclusion of any criterion will modify the final result of the classification exercise. Equally, the definition of the criteria is fundamental to avoid differing interpretations. For this reason, it is necessary to make explicit what is meant by each of the categories included in each criterion. The list of criteria runs as follows:

(1) **Physiognomy** corresponds to the growth forms dominant in the communities, which can be trees, shrubs, grasses, or non-vascular plants (see Table 3.1 in Appendix A.2). The dominance of these elements or attributes determines the physiognomy of the plant community: (**I**) forest (trees as the dominant element), (**II**) shrubland (shrubs predominating), (**III**) grassland (dominated by grasses), and (**IV**) non-vascular, for example in the case of algae or soil covered with another type of plant with negligible vascular systems (Tables 3.2, 3.3, 3.4, 3.5, 3.6, 3.7, 3.8 and 3.9).

(2) Climate as a second criterion conforms with two analyses: García (2004, p. 24) and Holdridge (1967). Both of their approaches were drawn upon for this present work. The first was the limitation to three climate types expressing thermal limits—(1) temperate (temperate and semi-cold), (2) tropical, for hot and semi-hot biomes, and (3) cold, for which García's originally proposed use is maintained. The temperate and tropical climates are divided into humid and dry, and the precipitation limits were proportioned from Holdridge's 1967 schematic, placing them in the corresponding climatic type. When a plant community develops naturally in at least three climate types, it is denominated

Table 3.1 SECLAVEMEX physiognomic criteria and their description

Criteria	Dominant attributes	Qualifiers definition and description of the dominant elements or attributes
Physiognomy	Forest	Dominated by woody stem plants above 5 m tall
Growth form	Shrubland	Dominated by plants with one of more woody or succulent stems. Usually of less than 5 m in height or arborescent or arbofrutescent plants over 5 m high
	Grassland	Dominated by plants with no woody base
	Non-vascular	Plants with limited or not vascular system

Table 3.2 SECLAVEMEX climatic criterion and its definition

Criteria	Dominant attributes	Qualifier's definition and description of the dominant elements or attributes
Climate	Humid temperate	Climates with a mean annual temperature between 5 and 18 °C and precipitation from 1,000 to >4,000 mm
	Dry temperate	Climates with a mean annual temperature between 5 and 18 °C and precipitation from 0 to 1,000 mm
	Humid tropical	Climates with a mean annual temperature above 18 °C and precipitation from 1,000 to >4,000 mm
	Dry tropical	Climates with a mean annual temperature above 18 °C and precipitation from 0 to 1,000 mm
	Cold	Type E (T) C (cold and temperate). Climates with mean annual temperatures between −2 and 5 °C and precipitation from 0 to 2,000 mm
	Euriclimatic	Cold, temperate, or tropical

Table 3.3 SECLAVEMEX physiognomic criteria and their description

Criteria	Dominant attributes	Qualifier's definition and description of the dominant elements or attributes
Phenology of the foliage	Deciduous	Plant communities where 75–100 % of the upper stratum individuals lose their leaves during the dry season
	Subdeciduous	Plant communities where 50–75 % of the upper stratum individuals lose their leaves
	Evergreen	Plant communities where 75–100 % of the superior-canopy individuals remain green year round
	Subevergreen	Plant communities where 25–50 % of the upper stratum individuals lose their foliage during the dry season

a euriclimatic climate (Table 3.1). The —climate‖ criterion, more than being an ecological differentiation factor, is a framework of nomenclature relevancy (Table 3.2).

(3) **The phenology** of foliage refers to the permanence or dropping of leaves at a determined time of the year by the individual-dominating plant communities

Table 3.4 The presence of spines and vertical spikes and their SECLAVEMEX definition

Criteria	Dominant attributes	Qualifiers definition and description of the dominant elements or attributes
Presence of spines or vertical spikes	Spiny	Plant community dominated by individuals with spines
	Spineless	Plant community dominated by individuals without spines
	Subspineless	Plant communities with a density superior to 30 % of individuals with spines but inferior to 70 %

Table 3.5 Leaf characteristics and their SECLAVEMEX definition

Criteria	Dominant attributes	Qualifier's definition and description of the dominant elements or attributes
Leaf characteristics	Needle-leaved	Needle-shaped leaf
	Huddler	A clustering of flexible sprouts over a small area.
	Angustifolius	Narrow, flatted leaf
	Caespitose	Turf leaf (developed through creeping stems)
	Scales-leaved	Scale-shaped leaf
	Broadleaved	Broad, flatted leaf
	Liniarifolia	Narrow leaf with parallel edges
	Megaphyllous	Leaf larger than $\pm 150\ cm^2$
	Microphyllous	Leaf smaller than $\pm 2.5\ cm^2$
	Rosetophylous	Clustered leaves, generally helically, forming a rosette
	Others	Various unconsidered types of leaves

Table 3.6 Succulence and its SECLAVEMEX definition

Criteria	Dominant attributes	Qualifier's definition and description of the dominant elements or attributes
Succulence	Succulent	Communities with a predominance of plants with succulent or fleshy stems, leaves, and roots
	Non-succulent	Community of non-succulent dominant plants, neither in leaves nor in stems

(in an area determined by its homogeneity). These have been split into four groups consisting the following communities: deciduous, subdeciduous, perennial, and subperennial (Table 3.3). Deciduous communities are those in which the dominant species in the upper canopy lose over 75 % of their leaves during the annual dry season. Perennials communities are considered such where over 75 % of species do not simultaneously lose leaves at any time of the year. The intermediate terms cover those situations over 50 % for subdeciduous and between 25 and 50 % for subperennial (Table 3.3).

Table 3.7 Floristic composition and its SECLAVEMEX definition

Criteria	Dominant attributes	Qualifier's definition and description of the dominant elements or attributes
Floristic composition	Family	Family is a mid-level taxon within the system of taxonomic categories containing one or more genera (The taxon "Family" can be dominant is some community for example herbs in grassland and Pinaceae in some temperate forests)
	Genus	The taxon "genus" can be dominant in some plant communities, for *Quercus* or *Pinus* in holm or pine forests
	Species	"Species" is the basic taxon with the hierarchical taxonomic system. Numerous plant communities can be discerned by the dominance of 1–3 or more species

Table 3.8 Substratum type and its SECLAVEMEX definition

Criteria	Dominant attributes	Qualifier's definition and description of the dominant elements or attributes
Substratum type	Gypsophile	Plant community adapted to gypsum rich soils
	Halophytic	Communities adapted to high salinity soils
	Aquatic and subaquatic	Plants adapted to water covering the soils
	Others	Other type of substrata which can be relevant

Table 3.9 Principal growth forms of organisms permitting the definition of physiognomic criteria for vegetation

Trees	A woody plant that generally has a single main stem and a more or less definite crown, or palms of arborescent appearance, mostly well above 5 m tall)
	Needle-leaved tree (mainly conifers, pine, spruce, larch, redwood, etc.)
	Broadleaved deciduous tree (leaves shed in the temperate zone winter or in the tropical dry season)
	Broadleaved evergreen tree (many tropical and subtropical trees, mostly with medium-sized leaves)
	Thorn tree (armed with spines, in many cases with compound, deciduous leaves, often reduced in size)
	Evergreen sclerophyllous tree (with smaller, tough, evergreen leaves)
	Succulent tree (primarily cacti and succulent euphorbs)
Shrubs	Shrubs (smaller woody plants, mostly below 5 m tall) or plants with arborescent or arbofrutescent appearance taller than 5 m (except palms above 5 m tall)
	Shrubs with needle-scale-shaped leaves (mainly conifers, junipers, taxus, etc.). Broadleaved deciduous shrub (leaves shed in the temperate zone winter or in the tropical dry season) Broadleaved evergreen shrub (many tropical and temperate shrubs, mostly with medium- to small-sized leaves)

(continued)

Table 3.9 (continued)

	Thorn shrub (armed with spines, in many cases with compound, deciduous leaves, often reduced in size)
	Evergreen sclerophyllous shrub (with smaller, tough, evergreen leaves), palm shrub (rosette shrubs, unbranched with a short crown of leaves)
	Dwarf shrub (low shrubs spreading near the ground surface, less than 50 cm high)
	Ferns (plant with arborescent form without branches and with a crown with an apical tuft of fronds bamboos (arborescent in appearance), subshrub (lianas or suffrutices with woody climbers or stem and high, dry branches during the dry season), succulents (cacti and certain euphorbiaceae or crassulaceae)
Herbaceous	Herbs (plants without perennial aboveground woody stems)
	Forb (herbs other than ferns and graminoids)
	Graminoid (grasses, sedges, and other grass-like plants)
	Fern (pteridophytes, ferns, clubmosses, horsetails, etc.)
	Succulent forb
	Aquatic plants (floating, emergent, or submerged), Epiphytes or parasites (plants growing complementarily on other plants)
	Climbers o vines, other unrecognized or unvalued forbs
Non-vascular	Small plants with limited or non-vascular system growing on soil or rock surfaces.
	Moss
	Liverwort/hornwort
	Lichen
	Alga

Modified from Jennings et al. (2002)

(4) ***The presence of spines and vertical spikes***. These are axial or appendicular parts hardened, sharp, or pointed present on some plants (on branches, trunks, or leaves). Plant communities can present dominance of spiny individuals (spiny communities), of spineless individuals (spineless communities) or with a relatively equal combination of spiny and spineless individuals (subspineless communities (Table 3.4).

(5) **Leaf characteristics**. Ten main leaf types have been chosen for their form, size, position, and appearance while developing (modifications on Font Quer 1982 and Raunkiær 1934). These characteristics represent different relevancies in plant communities. The option —others‖ is an attempt to cover those communities where the principal leaf types are not considered in the list provided (Table 3.5).

(6) **Succulence** refers to plants with fleshy and juicy tissues. They are distinguished from communities of plants with succulent steams (Crassicaules), some species of the families apocynaceae, asclepiadceae, cactaceae, and euphorbiaceae along with those with succulent leaves (crassiphollous) and some species from the families agavaceae, aizoaceae, crassulaceae, liliaceae, and portulacaceae (Table 3.6).

(7) **Floristic composition** includes three hierarchical levels from the taxonomic system for plants: (**a**) the family level, where reference is made to knowledge possessed about the plant community and the dominance of determined plants (for example, *graminaceae* or *poaceae*), (**b**) the dominant genus level of determined vegetation units, such as the case of the *Quercus* or *Pinus* forests, and (**c**) the dominant species level in a determined plant community (numerous plant communities can be distinguished by the dominance of one to three or more plant species. It is essential to carry out fieldwork or post-fieldwork activities with, at the least, support from experts in floristics and taxonomy to be certain as to the demands of this criterion (Table 3.7).
(8) **The criterion** substratum type refers to soils exercising a markèd influence on the presence of determined plant species which tolerate certain conditions, such as the abundance of gypsum (calcium sulphate) or of elevated salinity. Other conditions to be considered include important characteristics such as flooding frequency and the abundance of rocks (Table 3.8).

Secondary Vegetation

In essence, it is recognized that the majority of current vegetation types are an expression, sometimes totally, of the appropriation of the earth by humanity. Even so, from a predominantly ecological–functional perspective, vegetation is maintained through long-term evolutionary processes, which exceed by a long chalk the actions of man, such activity being restricted to only a few millennia (Ehrlich and Ehrlich 1992). During the last two centuries, however, this activity has had unimaginable repercussions, and it is for this reason that it is necessary to recognize a level in which the type of expression adopted as a result of human activity is explicit. It is obvious that human activity is not detected exclusively at a single level. In fact, all vegetation types, independently of the type of level or category to which they belong, can demonstrate an expression of disturbance or degree of relegation to secondary status. The vegetation classification process based on human activity leads to —land use typology‖, a topic which, due to its complexity, occupies a separate sphere.

Two of the most widely used criteria to define vegetation types in Mexico were excluded, given that they were dispensable and were also, to a certain degree, ambiguous. One refers to the relative height of the dominant physiognomic–climatic biotope. Types such as —Selva Baja and Mediana y Alta‖ (—Low, Mid, and High Jungle‖), for example, are only used for tropical forests, and there is no reason to link them to the remaining biotopes. Their inclusion, however, would be resorted to in defining criteria of relative heights for each biotope, varying from place to place and which, in a large part, are reported as per the perception of the author carrying out the vegetation survey. An exercise in attempting to reduce ambiguity in terms of denotations of vegetation heights led to Table 3.10, which was designed

Table 3.10 Definition of the height criteria for plant community strata in a vertical line

Stratum height (optional criterion)	High	If it is a forest with a canopy higher than 30 m, if it is a shrubland with a canopy of 3 > 5 m, or if it is grassland with a canopy higher than 2 m
	Mid	If it is a forest with a canopy between 15 and 30 m, if it is shrubland with a canopy between 1 and 3 m, or if it is grassland with a canopy between 0.5 and 2 m in height
	Low	If it is a forest with a canopy 5 and 15 m, if it is shrubland with a canopy lower than 1 m, or if it is grassland with a canopy lower than 0.5 m

Modified from (González-Medrano 2003)

to include height information, where it is considered necessary for a particular region, but applying it uniformly in all cases. It should be considered that, to be useful, this height criterion must be placed after the phenology attribute.

The other criterion excluded was the use of toponyms for the designation of a type of vegetation. Among those highlighted are —the Tamaulipeco matorral‖ (INEGI 1991) or pine and oak forest in the Sierra de Juárez (Dinerstein et al. 1995). It is clear that the systematic use of typonyms would aid in the understanding of distribution patterns, but they do not denote the botanical–geographical particularities of a vegetation type. For this reason, seeking to standardize them is not relevant to the classification system.

Hierarchical Levels in Secondary Vegetation

A system of eight nested hierarchical levels is being proposed derived in the main from proposals already in existence worldwide (Mueller-Dombois and Ellenberg 1974; Faber-Langendoen 2004). The first three levels denote predominantly physiognomic–environmental criteria. The next three levels are purely botanical–floristic in nature. Level VII is a mixture of characteristics of a floristic and substratum nature (Table 3.11). Additionally, it was fundamental to begin to coin nomenclature to designate each of the proposed levels. In this way, each level is associated with a name best describing it based on a review of the way it is used. A brief description was fundamental to assist the non-expert reader in understanding the level, its denomination, and its definition, as illustrated in Table 3.11, where the hierarchical structure of the eight levels, the denomination given to each, and the definition referring to the authors citing them can be seen.

Based on Tables 3.11 and 3.12, it was necessary to assign to each level criteria permitting the user to discriminate between the vegetation types belonging to each. It should, therefore, be clarified that, for each level, there will be a number of —vegetation types‖, a term prevalent in published research, to denote varied forms of vegetation grouping. In this way, the levels and the criteria best typifying each of

Table 3.11 Terms used to denominate different hierarchical levels in the proposal

Denomination	Definition
Bioma	Characterized by the physiognomic uniformity of mesoclimatic vegetation and includes the body of associated fauna; possesses a characteristic biotype constitution (Shelford and Olson 1935; Carpenter 1939)
Major formation	Term used by Zamudio et al. (1992: 13) to indicate a higher level or vegetation type (Formation) determined by climatic conditions *sensu lato* and the dominant life form
Formation	Used in this work as synonymous with a vegetation type of defined phenology and structure and characterized by its physiognomic structural uniformity
Subformation	Determined by botanical characteristics such as the presence of spines, the leaf morphology, and succulence
Series of associations	Determined floristically, and the taxon characterizing it can be a family or genus, meaning that it is a set of associations forming a series of plant communities which vary continuously
Association	The 1910 Brussels Congress defined it as "a plant community of determined floristic composition, a unit of seasonal conditions and a physiognomic unit; it is the fundamental synecological unit." Here, the association is an easily recognized plant grouping, which must comprise dominant species
Subassociation	A clearly specific grouping defined by a dominant species, indicating specific substrata characteristics (salinity, chalky, aquatic, or subaquatic for example)
Facies	The condition or degree of secondarization of vegetation. Succession or conversion

them are shown in Table 3.12. This table sums up the levels, criteria, and characteristics required to describe each vegetation type.

By way of example, if a site is located within a mountainous framework, at an elevation over 2,500 m, the surrounding vegetation, especially the original will be the first to be described, and, if trees are dominant, the level I (Bioma) will indicate that this vegetation must be denominated forest. In the same way, it is possible to move on to Level II (major formation), if the vegetation is identified as being in a humid temperate climate (humid temperate forest) using information from a climatic map (isohyets, isotherms). If, moreover, a seasonal permanence in terms of the foliage can be distinguished, the community can then be assigned, a level III typification (formation), as in humid temperate evergreen forest, and, if said community is dominated by individuals with needle-shaped leaves (acicular), it could then be given a level IV typification (subformation) as in humid temperate evergreen needle-leaved forest. In this example, the characteristics of spines and succulence were omitted, given that the vegetation in question does not present these attributes. Additionally, it could be that the vegetation referred to is dominated by trees of the genus *Pinus*, which would permit a greater level of typification to be

Table 3.12 Table summarizing levels, criteria, and characteristics necessary for describing a vegetation type

Level	I	II	III	IV			V	VI	VII		VIII
Category	Bioma	Major formation	Formation	Subformation			Series of associations	Association	Subassociation		Facia
Criterion	Growth form	Climate	Phenology	Presence of spines	Characteristics leaves	Succulence	Floristic	Floristic	Floristic	Substratum	Condition
Characteristics dominant of community	Forest	Cold	Deciduous	Spiny	Needle-leaved	Succulent	Family or Genus dominant	Two or more dominant/unique species	Dominant/exclusive species	Gypsophilous	Conversion
	Shrubland	Humid temperate	Subdeciduous	Spineless	Clustered	Non-succulent		Preferential	Preferential	Halophytic	Succession
	Grassland	Dry temperate	Evergreen	Sub-spineless	Angustifolia					Others	
	Non-vascular	Humid tropical	Subevergreen		Caespitose						
		Dry tropical			Scale-leaf						
					Broadleaved						
					Linearifolia						
					Megaphyllous						
					Microphyllous						
					Rosetepholius						
					Others						

accorded—Level V: series of associations. If, moreover, a recognition of the characteristic species of pines[1] is achieved, then level VI can be reached. If the species are *Pinus psedostrobus* and *P. herrerae*, the name coined for this type of vegetation would be —"**humid temperate evergreen needle-leaved *Pinus psedostrobus-P. herrerae* Forest.**"

From this example, it follows that all those vegetation types distinguished by the same name fit within a group presenting better affinity within itself than with any other type of vegetation groups or other category. A look at various works describing the vegetation in certain places with a view to organizing them according to the proposed system in Table 3.12 illustrates this further. The description of each case study pursues three aims. One is to show SECLAVEMEX's versatility in terms of including previous works. The second is to illustrate the manner in which it disaggregates the most common inconsistencies. The third is to orient the reader towards a process of inferring the criteria to designate a —vegetation type‖ based on a given description. To illustrate even further some of SECLAVEMEX's operational elements, nine case studies are used, which permit the comparison of related vegetation types and which cover a wide range of Bioma (Appendix A.3).

The first case (Appendix A.3): A vegetation type described by Martínez and Galindo-Leal (2002) corresponds to a Level I (or Bioma) forest, given that there is a clear dominance of trees. For Level II (major formation), humid tropical climate is inferred from temperatures above 18 °C with precipitation above 1,000 mm. Level III (formation) corresponds with —deciduous‖ because of the dominance of leaf fall. For Level IV (subformation), —broadleaved‖ is inferred from the dominant leaf morphology. The authors give *Bursera* as one of the dominant genera, meaning it is used to designate Level V (series of associations). Within this vegetation type, the authors distinguish between nine primary —associations‖. For this example, only one association is referred to, which is defined by the presence of *Pseudobombax elliptucum-Bursera simaruba*, permitting advancement to level VII (association). Finally, and derived from the authors' description, it follows that *Lysiloma latisiliqua* forms —secondary origin associations‖. The nomenclature coined to refer to this type of vegetation would be —"**Humid tropical broadleaf deciduous *Pseudobombax ellipticum-Bursera simaruba* forest with a *Lysiloma latisiliqua* subassociation.**"

Finally, Table 3.13 uses the original nomenclature given for each of the case studies illustrated in Table 3.12, and, where they might be considered to be in the same hierarchical level, nomenclature coined as per SECLAVEMEX has been assigned them.

By way of conclusions, it should be noted that SECLAVEMEX typifies the number of vegetation types precisely, both in describing and in denominating them. The example described in Table 3.12 serves to consolidate the typification. Another example (Table 3.14), developed for a vegetation study in a particular region (Velázquez et al. 2010b), demonstrates a series of benefits namely:

[1]Characteristic species are those which contribute to the vegetation's appearance or physiognomy.

Table 3.13 Comparison of the results from the application of the SECLAVEMEX to the examples in Appendix A.3

Previous vegetation type name	Case study author	Vegetation type name according to SECLAVEMEX (Level IV)
Low deciduous jungle	Martínez and Galindo-Leal (2002)	Humid tropical broadleaved deciduous forest
	Durán (2004)	Dry tropical broadleaved deciduous forest
Quercus forest	González-Elizondo et al. (1993)	Dry tropical broadleaved deciduous *Quercus* forest
Holm forest	Ramírez-Marcial et al. (1998)	Humid tropical broadleaved deciduous forest
Mountain mesophyll forest	Luna et al. (2001)	Dry tropical broadleaved evergreen forest
	Luna et al. (2001)	Humid temperate broadleaved evergreen forest
Spiny shrub	Romero and Rojas (1991)	Dry temperate spiny deciduous Shrubland
Xerophile shrub	Pérez-García et al. (2001)	Humid tropical succulent evergreen shrubland
Grassland	Pérez-Calix (1996)	Humid temperate angustifolia perennial herbaceous
	Briones (1991)	Dry tropical angustifolia deciduous herbaceous

I. The number of vegetation types within a territory, given that it will depend solely on the diversity of geobotanical expression and will cease to depend on the author of the system used.
II. In each level, SECLAVAMEX permits differentiation between vegetation types based exclusively on the geobotanical characteristics, which permit (dis) aggregation between types of greater or lesser affinity based on their attributes —for example, broadleaved deciduous dry tropical forests which are home to both the genus *Cordia* (traditionally recognized as deciduous tropical forest or low deciduous jungle) and others of the genus *Quercus* with deciduous species (traditionally included as holm forests associated with temperate environments with perennial phenology).
III. Within SECLAVEMEX, and unlike in the majority of previous systems, the mesoclimatic characteristic is used rigorously to distinguish between vegetation types in each level. For example, in Appendix A.3, within the *Low Deciduous Jungle* category, there are two groupings, one being humid tropical and the other dry tropical. The implications at lower levels are multiple, given that discernment is possible between types which could conventionally be considered similar or equivalent.
IV. The name, although extensive, reflects a synthesis of the particular characteristics of each vegetation type.

Table 3.14 Example of terminology used to describe different types of vegetation in the Sierra Ajusco-Chichinautzin (Velázquez et al. 2010b)

Hierarchical Level	Hierarchical SECLAVEMEX
Coniferous forest	Humid temperate needle-scale liniarifoliate evergreen forest
Pinus hartwegii forest	*Pinus Hastwegii* humid temperate needle-leaved evergreen forest
Pinus forest	*Pinus* humid temperate needle-leaved evergreen forest
Abies religiosa forest	*Abies religiosa* humid temperate liniarifoliate evergreen forest
Mixed forest	Humid temperate needle–broadleaved evergreen–subevergreen forest
Pinus spp.-*Quercus* spp. forest	*Pinus* and *Quercus* humid temperate needle–broadleaved evergreen forest
Pinus spp.-*Alnus* spp. forest	*Pinus* and *Alnus* humid temperate needle–broadleaved evergreen–deciduous forest
Holm forest	*Quercus* humid or dry temperate broadleaved subevergreen forest
Quercus crassipes and *Quercus obtusata* forest	*Quercus crassipes* and *Q. obtusata* humid temperate broadleaved subevergreen forest
Quercus rugosa forest	*Quercus* humid temperate broadleaved subevergreen forest
Quercus mexicana and *Quercus crassipes* forest	*Quercus* humid temperate broadleaved subevergreen forest
Quercus laurina forest	*Quercus* humid temperate broadleaved subevergreen forest
Mountain mesophyll forest	Humid temperate broadleaved evergreen forest
Xerophile shrubland	Dry temperate shrubland
Quercus shrubland	Dry temperate broadleaved forest
Senecio preacox shrubland	*Senecio preacox* dry temperate broadleaved deciduous shrubland
Hechtia podantha shrubland	*Hechtia podantha* humid or dry temperate rosetophilous evergreen shrubland
Bursera shrubland	*Bursera* humid temperate broadleaved deciduous forest
Salix shrubland	*Salix* humid temperate angustifolia evergreen grassland
Other vegetation types	
Mega-rosetophilous	Dry temperate rosetophilous evergreen shrubland
Alnus forest	*Alnus* humid temperate broadleaved deciduous forest
Grassland	Herbaceous
Subalpine grassland	Humid temperate herbaceous
grassland	Humid temperate broadleaved perennial herbaceous
Aquatic and subaquatic vegetation	
Aquatic communities	Humid temperate broadleaved deciduous aquatic herbaceous
Subaquatic communities	Humid temperate broadleaved deciduous subaquatic herbaceous

Reproduced with the permission of Investigación Ambiental

Use of Abbreviations in Vegetation-Type Names

With the aim of facilitating the expression of nomenclature used in the plant categories, it is possible to make use of abbreviations. The abbreviation recommended for long names (which would not lose their validity) would be used solely to express a category quickly and would consist of the name of the element of the first level and the name of the element(s) in the last level—for example, vegetation types in Appendix A.3: —broadleaved forest‖, —Quercus forest‖, —Liniari-latifolia forest‖, —spiny shrubland‖, —Succulent shrubland‖, and —angustifolia grassland‖, corresponding with key nomenclature, which, in some cases, will be the vernacular name.

Mexican Vegetation Types Based on SECLAVEMEX

This section commences the task of describing, distinguishing between, ranking, and naming the vegetation types of Mexico in their various levels of organization. Tables 3.10, 3.11, and 3.12 (see them in Appendices A.4, A.5, and A.6) show the categories or vegetation types recognized for Mexico based on SECLAVEMEX as far as a Level VI (series of associations) and illustrates the complexity of vegetation types at each level based on the findings of the various authors considered. In a preliminary manner, it can be seen that passing from a higher level to the level immediately below leads to an increased representation of diversity.

(1) Bioma: 4 units
(2) Major Formations: 15 units
(3) Formations: 28 units
(4) Subformations: 58 units
(5) Series of Associations: 134 units

The first category (Bioma) includes the lower categories in the same way that each successively subsequent level includes those below itself. Such is the nested character of the proposal. In the Bioma (forests), there are 56 series of associations recognized; shrublands have 50, grasslands 26, and non-vasculars 2; this last vegetation unit being little known or taken into account by vascular plant or vegetation researchers. In this classification system, the inclusion of climate criteria allows for more precise knowledge as to the vegetation units being analysed, one example being humid tropical acicular perennial forests (tropical zone pine forest), a denomination previously related to temperate climates.

The theme of diversity in plant communities or beta-diversity (the rate of species replacement from one community to another) is still a long way from being exhausted. Based on the project plan described for SECLAVEMEX, it is hoped that any research with the objective of studying vegetation can be described ranked, named, and thus contextualized. Beyond this point, in a few years, it should be possible to depend on a complete and robust inventory of vegetation types from Biomas all the way through associations. In the following chapter, SECLAVEMEX's inclusive nature is highlighted, and a general overview of the proposed system is provided with respect to other macroregional and global projects.

References

Briones O (1991) Sobre la flora, vegetación and fitogeografía de la Sierra de San Carlos, Tamaulipas. Acta Botánica Mexicana. Instituto de Ecología A.C. Pátzcuaro, Michoacán México, 16: 15–43.

Carpenter JR (1939) The Biome. Amer. Midl. Nat. 21: 75–91.

Challenger A (1998) Utilización and conservación de los ecosistemas terrestres de México. Pasado, Presente and Futuro. Comisión Nacional para el Conocimiento and Uso de la Biodiversidad - Instituto de Biología. Universidad Nacional Autónoma de México - Agrupación Sierra Madre, A.C. México, D.F.

Dinerstein E, Olson D. Graham D, Webster A, Primm S, Bookbinder M, Ledec C (1995) Una evaluación del estado de las eco-regiones terrestres de América Latina and el Caribe. Banco Mundial, WWF. Washington D.C.

Don_Faber-Langendoen (2004) HRWG: Compilation of Issues and Decisions. Microsoft Words document.

Durán ME (2004) Estructura, diversidad and mortalidad del componente arbóreo en un mosaico ambiental de Chamela, México. Tesis de Doctorado. Universidad Nacional Autónoma de México. México, D. F.

Ehrlich PR, Ehrlich AH (1992) The value of biodiversity. Ambio 21: 219–226.

Ferrusquía-Villafranca I (1998) Geología de México: una sinopsis. In: Ramamoorthy TP, Bye R and Lot A 1998. Diversidad biológica de México: orígenes and distribución. Instituto de Ecología. Universidad Nacional Autónoma de México.

Font Quer P (1982) Diccionario de Botánica. Labor S.A., Barcelona.

García E (2004) Modificaciones al Sistema de Clasificación Climática de Köppen. Instituto de Geografía. Universidad Nacional Autónoma de México. Serie Libros, 6. México, D.F.

González-Medrano F (2003). Las comunidades vegetales de México. Propuesta para la unificación de la clasificación y nomenclatura de la vegetación de México. Instituto Nacional de Ecología, México, DF.

González-Elizondo S, González-Elizondo M, A Cortés O (1993) Vegetación de la Reserva de la Biosfera - La Michilía, Durango, México. Acta Botánica Méxicana. Instituto de Ecología, A.C. Pátzcuaro, Michoacán. México. 22: 1–104.

Holdridge LR (1967) Life zone ecology, San José, Costa Rica.

INEGI (Instituto Nacional de Estadística Geografía e Informática) (1991) Datos básicos de la Geografía de México.

Jennings M, Loucks O, Glenn-Lewin D, Peet R, Faber-Langendoen D, Grossman D, Damman A, Barbour M, Pfister R, Walker M, Talbot S, Walker J, Hartshorn G, Waggoner G, Abrams M, Hill A, Roberts D, Tart D. (2002) Standards for associations and alliances of the U.S. national vegetation classification. Version 1.0. Vegetation Classification Panel, Ecological Society of America. Washington, D.C. USA.

Levin SA (ed) (2009) The Princeton Guide to Ecology. Princeton University Press, Princeton.

Luna I, Velázquez A, Velázquez E (2001) México. In: Kappelle, M. and A. D. Brown. Bosques nublados del neotrópico. Instituto Nacional de Biodiversidad. Fundación Agroforestal de Argentina. Comité Holandés de la Unión Mundial para la Conservación de la Naturaleza.

Universidad de Amsterdam. Laboratorio de Investigaciones Ecológicas de Los Yungas de Argentina. Costa Rica: 183–229.

Martínez E, Galindo-Leal C (2002) La vegetación de Calakmul, Campeche, México: Clasificación, Descripción and Distribución. Bol. Soc. Bot. México 71: 7–32.

Mueller-Dombois D. Ellenberg H (1974) Aims and Methods of Vegetation Ecology. John Wiley & Sons, Manchester.

Pérez-Calix E (1996) Flora and vegetación de la cuenca del Lago de Zirahuén, Michoacán, México. Flora del Bajío and de Regiones Adyacentes. Fascículo Complementario XIII. Instituto de Ecología. Pátzcuaro, Michoacán. México.

Pérez-García EA, Meave J, Gallardo C (2001) Vegetación and flora de la región de Nizanda, Istmo de Tehuantepec, Oaxaca, México. Acta Botánica Mexicana. Instituto de Ecología A.C. Pátzcuaro, Michoacán. México, 56: 19–88.

Ramírez-Marcial N, Ochoa-Gaona S, González-Espinosa M, Quintana-Ascencio PF (1998) Análisis florístico and sucesional en la estación biológica Cerro Huitepec, Chiapas, México. Acta botánica Méxicana. Instituto de Ecología A.C. Pátzcuaro, Michoacán. México, 44: 59–85.

Raunkaier C (1934) The life forms of plants and statistical plant geography. Clarendon, Oxford.

Romero S, Rojas C (1991) Estudio florístico de la región de Huehuetoca, Estado de México. Acta Bot. Mex. 14:33-58.

Rzedowski J (1978) Vegetación de México. Limusa. México, D.F.

Rzedowski J (1991a) Diversidad and orígenes de la flora fanerogámica de México. Acta Botánica Méxicana. Instituto de Ecología A.C. Pátzcuaro, Michoacán, México, 14: 3–21.

Rzedowski J (1991b) El endemismo en la flora fanerogámica mexicana: una apreciación analítica preliminar. Acta Botánica Mexicana. Instituto de Ecología A.C. Pátzcuaro, Michoacán, México. 15: 47–64.

Rzedowski J, Medina L, Calderón de Rzedowski G (2005) Inventario del conocimiento taxonómico, así como de la diversidad and del endemismo regionales de las especies mexicanas de Bursera (Burseraceae). Acta Botanica Mexicana 70: 85–111.

Shelford VE, Olson S (1935) Sere, climax and influent animals with special reference to the transcontinental coniferous forest of North America. Ecology. 16: 375–402.

Velázquez A, Medina C, Reygadas D (2010b) Síntesis geobotánica de las sierras Ajusco-Chichinautzin. Investigación ambiental. Ciencia and política pública, Instituto Nacional de Ecología-SEMARNAT. 2(1) http://www.revista.ine.gob.mx/article/view/90.

Wendt T (1998) Composición, afinidades florísticas y orígenes de la flora arbórea del dosel de los bosques tropicales húmedos de la vertiente mexicana del Atlántico. In Diversidad Biológica de México: orígenes y distribución, Ramamoorthy T, Bye R, Lot A, Fa J (comp.). Instituto de Biología, Universidad Nacional Autónoma de México. México, D. F.

Zamudio R, Rzedowski J, Carranza E, Calderón G (1992) La vegetación en el estado de Querétaro. Instituto de Ecología, A.C.- CONCYTEQ.

Chapter 4
The Inclusive Nature of SECLAVEMEX

Introduction

The objective of this chapter was to return to the classification systems used across the whole of Mexico to illustrate explicitly the mechanics of affinity and compatibility between these various systems and SECLAVEMEX, retaining the strengths of previously proposed systems but resolving inconsistencies prevalent among them. In this way, after analysing the classification systems most widely used in Mexico and how they correspond to SECLAVEMEX, the operative process for including a new type of vegetation is illustrated pragmatically. The final section of this chapter describes the preparation, fieldwork, and post-fieldwork tasks necessary to carrying out this process successfully.

How SECLAVEMEX Corresponds with Previous Systems in Mexico

SECLAVEMEX's inclusive nature comes from systemizing existing knowledge on vegetation types in Mexico. To this end, the main part of the work has been to include all vegetation types reported in the four most prominent works on Mexico—those of Miranda and Hernández-X (1963), Rzedowski (1978), COTECOCA (1994) and González-Medrano (2003). Complementarily, also included are studies which, above and beyond presenting a classification system, lay out a legend for the cartographic expression UNESCO (1973), INEGI (1991), Dinerstein et al. (1995) and Palacios-Prieto (2000).

Each author's description for each type of vegetation in Mexico was carefully reviewed with a view to succinctness. The reasoning involved distinguishing, and on many occasions, further inferring the characteristics that an author had denoted to define a type of vegetation and what distinguished it from others. In this way,

A. Velázquez et al., *Standardized Hierarchical Vegetation Classification*,
Geobotany Studies, DOI 10.1007/978-3-319-41222-1_4

recognition for each vegetation type and the place corresponding to it within the SECLAVEMEX structure was achieved. It should be noted that, at times, it was necessary to extract characteristics not clearly described by the author with the result that there is the possibility that a vegetation type has been incorrectly placed. The task, though it might appear trivial, was not simple. By way of exemplification, below is provided a sample extracted from the primary natural vegetation types from each of the eight classification systems:

1. Miranda and Hernández-X (1963). Halophyte groupings formed filled the first level, Bioma with references all the way to the seventh level, subassociations (halophile), giving clear indications of the species set forming this grassland, namely those tolerant of high levels of salinity. It is possible to make inferences, but precise data on the constitution of the intermediate levels are missing.
2. UNESCO (1973). Open acicular sempervirent forest, which took up the first level, class (forest), the third, formation (evergreen), and the subformation (needle-leaved). While further description would be desirable, the lack of known characteristics makes it impossible to make specific inferences.
3. Rzedowski (1978). Abies forest, first level (forest). Fifth level, series of associations (Abies). The intervening levels can be inferred.
4. INEGI (1991). Low open forest. This denomination is found in the first level, Bioma (forest). The description refers to a pine-holm forest, structurally low with a high dominance of herbs. The herbaceous fits into the same category as the trees, but, without a doubt, it does not fit precisely into SECLAVEMEX.
5. COTECOCA (1994). Needle-leaved sclerophyllous. For this denomination, it can be inferred that it corresponds to the fifth level, series of associations, (Pinus and Quercus).
6. Dinerstein et al. (1995). Pinus and Quercus forest in the Sierra de Juárez. First level, Bioma (forest), fifth level, series of associations (Pinus and Quercus), the same as the preceding.
7. Palacios-Prieto et al. (2000). Pine forest. First level, Bioma, (forest). Fifth level, series of associations (*Pinus* spp.). The information is missing.
8. González-Medrano (2003). —Matorral innerme alto‖ comprised the first level—Bioma (shrubland) and the fourth level—subformation (spineless).

As can be seen, nearly all of the previous proposals have gaps in terms of the type of characteristics necessary to fill in every box in SECLAVEMEX. Even so, the majority of the descriptions are clear enough, particularly for those knowledgeable of vegetation. Exceptions, however, abound in that, in each place, it can be difficult to distinguish between types at their various grouping or hierarchical levels. In Table 4.1 (Appendix A.9), each of the vegetation types found in Mexico is described based on previously published research and its correspondence with the eight works reviewed. The result demonstrates a richness of vegetation types very much superior to that reported independently by each of the individual authors. It also shows that vegetation types originally considered to be homologous were not so comparable (see Table 4.1 in Appendix A.9).

Table 4.1 Levels and nomenclature in the different categories of the two classifications with the criteria adopted for each

Level	Nomenclature SECLAVEMEX	Criteria	Nomenclature Phytosociology	Criteria
I	Bioma	Physiognomy	Class	Physiognomy and floristic
II	Major formation	Temperature and humidity	Subclass	Bioclimate
III	Formation	Phenology		
IV	Subformation	Spines, leaf morphology	Order	Floristic composition
V	Series of associations	Floristic (dominance of Family or Genus)	Alliance	Vegetation series
				Floristic
VI	Association	Floristic (two or more dominant or indicative species)	Association	Floristic Possession of characteristic or indicative species
VII	Subassociation	Floristic (dominance of one species) and substratum	Subassociation	Floristic composition
VIII	Facies	Vegetation stadium	Facies	Floristic, successional stadia

The benefits of Table 4.1 are illustrated as follows: one is the (dis)aggregation of vegetation types which present certain similarities or distinctive characteristics offering the possibility of being characterized according to determined common elements, in their turn giving rise to perfectly defined groups corresponding to the vegetation types described by each author in higher, previously unconsidered levels. These hierarchical levels are crucial when dealing with various levels of biological organization, especially for the purposes of administration or natural-resource management. Another benefit is the possibility of including systematically new vegetation types not reported in the eight previous works and eventually of systemizing any work on vegetation new or old within the system proposed, as illustrated in Chap. 3. Finally, it is recognized that, from the systematic use of SECLAVEMEX's structure, it will be possible to include information originating in rigorously carried out regional studies.

Correspondence Between SECLAVEMEX and Current International Systems

Relationship with the Phytosociological Approach

The phytosociological approach (PA) is without a doubt the method most widely used at the global level, where the aim is to construct vegetation classification systems. Its hierarchical character and its rules and criteria denoting each vegetation unit are rigorous, on the one hand, and practical on the other, leaving it without equal at the global level (Werger 1974). The phytosociological approach has remained prevalent in Europe and in a good part of Asia and Africa for over a century, leading to the data bases now available, which are based on hundreds of thousands of comparable relevés, being thus a detailed inventory of diverse plant communities. In terms of comparisons between SECLAVEMEX and the PA, comparisons of both the structure and nature of the criteria are looked at. The (dis) similarities are presented in synthetic form in Table 4.1.

1. The SECLAVEMEX Biomas level and that of class in the PA are two similar concepts denoting the upper hierarchical level of an ecosystem. The first refers to a biotic system characterized, speaking in terms of the vegetation, by the dominant structural life form. The class is determined by the dominant plants' form of growth, but the name is generated from the species best describing or characterizing the condition.
2. In Level II, major formation and subclass are conventional terms. The two are related to the climate. The first refers to the temperature and humidity (humid temperate, dry temperate, humid tropical, dry tropical, and cold), and the second equally refers to elements bearing on humidity—pluvial, pluviseasonal, and xeric—combined with the temperature—that is the bioclimatic index. In the same way, its difference lies in the specific denotation made in the subclass.
3. The formation level is proposed for the two approaches, and, in both, it will use phenological criteria such as the differentiating criterion and characteristics of the strata height as a descriptive attribute.
4. The subformation level for SECLAVEMEX and order for the PA are the intermediate levels related to the broad physiognomic criteria and the sum of the characteristics of the higher levels. For SECLAVEMEX, the subformation includes geobotanical characteristics (succulence, leaf morphology, presence of spines) to denote the vegetation type. For the PA, this level, order, is defined by two or three characteristic species which group together in sets of alliances (lower intermediate level).
5. Series of associations for SECLAVEMEX and alliance for the PA are similar terms referring to a clearly dominant vegetation type by a genus or family taxonomic, indicating that they are a set of associations. In the PA, the alliance is the phytosociological unit which characterizes a grouping of associations of floristic composition, stadium, biotope and similar habitat. In each case, the criteria are floristic, but the denomination in the PA is based on species, while in SECLAVEMEX, it can be based on dominant families and genera.
6. Level VI. The term _association" in SECLAVEMEX refers to the predomination or abundance of two or more species best denoting the condition or habitat of a vegetation type. In the PA, _association" is the basic working unit, and one is differentiated from the other by the possession/existence of determined characteristics and differential species. The definition of _association" is, no doubt, rigorous in the PA and implies a profound knowledge of the flora.
7. Level VII, the subassociation, is determined by a generally dominant species or set thereof as a response to a type or condition of substratum atypical to the

regional climate (mesoclimatic) or of the relief or soil (mesomorphoedaphic). In general, the subassociation can be considered azonal sensu lato. For the PA, subassociation derives from the division of an association presenting certain variations in its floristic composition.

8. The final level, facies, is defined by the stadium or condition of the vegetation. For SECLAVEMEX, there arise two clear tendencies: one dealing with reversible alteration processes and thus potentially open to potential resilience on the part of the native vegetation and the other concerned with irreversible alteration to native vegetation because of the sheer scale of change, thus effecting a transformation, be it because of a cultural system or of a degree of secondarization for the vegetation with dominance by species native to the area. For the PA, the facies refer to variation in abundance and domination of species characteristic of a (sub)association. In both cases, anthropogenic action is the principal defining criterion of the condition.

Levels V, VI, VII and VIII, even if defined based on similar criteria, differ, in both approaches, greatly in their operative frameworks, a product of the nature of the vegetation in a country like Mexico, which has a high rate of endemism (in some regions, every third species is native) and is a centre of speciation in terms of taxonomic groups (e.g. holms, *Bursera,* and pines, among hundreds of others beyond plant species. Additionally, Mexico's flora has largely co-evolved in close relationship with human groups for over 6,000 thousand years, which makes it one of the richest domestication centres in the world (Ramamoorthy et al. 1998), all of which complicates the analysis, within the PA, of vegetation with strict criteria. The PA was, no doubt, devised to systemize knowledge on vegetation from places with prevalent conditions of mesoclimatic homogeneity where azonal conditions are exceptions in the form of few and well-defined ecotones with clear species dominance—particularly in those regions where there exists a complete and rigorous floristic inventory something not possible across most of Mexico, where 80 % land cover is transitional with diffuse limits between geological, lithological, geomorphological, edaphic, climatic, biogeographic, and thus ecological units. Exceptions abound, and, on occasion, they pose difficulties vis-à-vis the mesozonal connotation of a vegetation type (e.g the mountain mesophyll forest sensu Rzedowski 1978). There is, doubtless, much information of high scientific value which can currently be used to develop a system arranged along other levels of biological organization apart from the specific. As phytoecological knowledge deepens, it will be possible to improve the system proposed in this work.

Relationship Between SECLAVEMEX and International Proposals

The Federal Geographic Data Committee (FGDC) has been formed with the purpose of developing and integrating a vegetation classification system based on physiognomic criteria, floristics, and their respective nomenclature. Its aim is to organize regionally and across the continent the application of floristic methods of vegetation classification. The intention of said committee is not to arrive at a single vegetation classification system, but rather to propose systems which maintain a consensual platform aimed at facilitating communication between studies on the subject and to define conservation priorities (FGDC 2006). Canada and the USA have set themselves the goal of developing a system with the same naturalness and complexity as can be found in the PA. The advances in the structure of said project in 2008 are presented schematically in Table 4.2. In Table 4.3, the SECLAVEMEX proposal has been added along with how it corresponds with the phytosociological approach.

A clear correspondence is noted in Table 4.2 between the three approaches in their aggregation levels. The differences lie in the prevalent criteria in each level. The main effects are noted in the lower levels, but it is here where botanical criteria dominate the three systems.

Table 4.2 Levels, nomenclature, and selected criteria for vegetation classification in North America (FGDC 2008)

Taxonomic level	Criteria (plant adaptations)
Upper: Physiognomy plays a predominant role	
L1—Formation Class	Broad combinations of general growth forms (mesomorphic, xeromorphic, cryomorphic, lithomorphic, hydromorphic) that reflect basic moisture/temperature regimes
L2—Formation Subclass	Combinations of general growth forms that reflect global climatic factors
L3—Formation	Combinations of general and specific growth forms and physiognomy that reflect geographically widespread (global) topographic and edaphic factors, including broad altitudinal gradients
Middle: Both floristics and physiognomy play a significant role	
L4—Division	Combinations of specific growth formas and diagnostic plant taxa reflecting continental- or broad geographic-scale factors. Many character taxa are expected
L5—Macrogroup	A common set of growth forms and many diagnostic plant taxa (including some character taxa of the dominant growth forms), prefeentially sharing a broadly similar geographic range, mesoregional climate, and disturbance factors. In the case of semi-natural vegetation, large-scale anthropogenic modifications of these factors may occur. Many character taxa are expected
L6—Group	A common set of growth forms and diagnostic species (taxa) (including species of the dominant growth forms), preferentially sharing a similar set of edaphic, topographic, and disturbance factors, prevailing at a regional level
Lower: Floristics plays a predominant role	
L7—Alliance	Specific diagnostic species, including those from the dominant growth forms, with a moderately homogenous physiognomy and structure, which together reflect moisture, fertility, and disturbance gradients within a region. Typically contains at least one character (or strong differential) species
L8—Association	Specific diagnostic species or combination of species from any growth form, with an overall moderately to strongly homogeneous physiognomy and structure, which together reflect specific moisture, fertility and disturbance gradients within a region or landscape

Table 4.3 Relationship of the SECLAVEMEX proposal to the FGDC and phytosociological systems

Hierarchical levels, nomenclature and criteria						
Level	Standardized Hierarchical Classification System for Vegetation in Mexico (SECLAVEMEX)		Federal Geographic Data Committee (FGDC) (predominant in Canada and the USA)		Phytosociological approach (EF) (predominant in European Union)	
	Nomenclature	Criteria	Nomenclature	Criteria	Nomenclature	Criteria
I	**Bioma/**Class	Dominant growth forms	Formation Class	Dominant growth forms, moisture/temperature regimes	Class	Physiognomy and floristic
II	**Major Formation/** Subclass	Climatic	Formation Subclass	Physiognomy, and growth forms, mayor climatic conditions	Subclass	Bioclimatic
III	**Formation**	Phenology	Formation	Physiognomy and growth forms, topografic and edaphic conditions		
IV	**Subformation/** Order	Physiognomy/floristic	Division	Broad geographic area, dominant growth forms and diagnostic species	Order	Species composition
V	**Series of Associations/** Alliance	Floristic	Macrogroup	Physiognomy, biogeographic region, diagnostic taxa, regional climate and disturbance	Alliance	Series of association Species composition
VI	**Association/** Association	Floristic	Group	Growth forms and diagnostic species, regional edaphic, topographic and disturbance factors	Association	Species composition and diagnostic species
VII	**Subassociation/** Subassociation	Floristic and substratum	Alliance	Species composition, habitat, physiognomy and diagnostic species	Subassociation	Species composition and diagnostic species
VIII	**Facie/**Facie	Successional stage/Floristic	Association	Species composition, diagnostic species, habitat and physiognomy	Facies	Floristic and successional stage

In Mexico, there is still a long way to go before there will be a complete phytoecological inventory. As the various thousands of regional and local vegetation studies are capitalized on and are transferred rigorously into a system such as SECLAVEMEX offers, it will be possible in record time to compile said inventory.

Operative Strategies of the Proposed System

Pre- and Post-Fieldwork Research Phases

Recapitulating, SECLAVEMEX assumes that —*vegetation*‖ has diverse expressions, which can be distinguished between via discriminatory attributes in each of the eight proposed levels. In this way, for its systematic application, it is necessary for the —*subject*‖ (researcher, student, technician, or other) responsible for the development of the vegetation study to collect information from a wide range of previous field studies. In the case of Level I (Physiognomy), it will be relevant to understand minutely that attributes characterize and differentiate between a forest, shrubland, grassland, et cetera. The physiognomic aspect consists of describing the structure, which is the basic organization and the measure of the form, and of the appearance of the vegetation (Whittaker 1973). The physiognomy considers six important characteristics of vegetation—namely species dominance, life forms, stratification, foliage density, cover, and spatial disposition. Species dominance is one of the most important criteria and can be expressed through its abundance, structural complexity, size or Bioma. Dominant species define the appearance of the vegetation, and thus, it becomes crucial to know what they are in order to be able to assign a dominant physiognomical type. The task is not trivial, for, often, it reveals a great number of particularities within a given territory. It should be remembered, therefore, that the study of vegetation pursues zonal ends and that particularities correspond to azonal situations. If the dominant physiognomy is a forest in the regional context, this fact does not exclude the presence of shrubland or herbazales (herbaceous or forbs) within, but the dominant condition would remain —forest‖. At this level, it is suggested that the use of qualifiers should be limited, as they tend to confuse a reader if they are not precisely defined—for example, the use of terms such as _open forests", _low forests", _dwarf forests", _microforests", among the many others abounding in published research, which terms, in general, represent local assessments and denote a relative condition, rendering the assessment subjective.

For Level II, it is necessary to make a priori a rigorous assessment of the regional climate where the vegetation study will be carried out. In this way, the temperature (cold, temperate, or tropical) and the precipitation (dry or humid) as listed in the different combinations reported for Mexico by García (2004) permit a dominant regional climatic condition to be assigned. One way of finding this type of information could be through CONABIO (http://conabioweb.conabio.gob.mx/metacarto/metadatos.pl), where the map can be consulted on the screen or downloaded in various formats. It should be explained that the scale of Dr. García's original work (1:500,000) does not allow for detailed inferences. Doubtless, in a regional vegetation study, there will be enclaves with plant species native to different climates, as the map shows. At this point, it should be taken into account that climate defines vegetation, but vegetation modulates climate, and thus, the vegetation itself is often used to coin a dominant climate type in particular enclaves. In the regional context, it will doubtless be the climate defining the conditions, with the native vegetation then resulting from them. The particularities of enclaves of vegetation contrasting with the regional (zonal) vegetation can spring from local hydrological, geopedological, or anthropogenic conditions known as hydro- or edaphic series, or, in their defect, from a history of particular use, among possible causes (anthroposeries).

For Level III (foliage condition or phenology), it becomes necessary to consider that the dominant species of the different units recognized present behaviours linked to the temporality of the foliage, the discrimination of which, on occasions, is not so evident. Common examples can be seen in transactional zones or ecotones where the dominance of deciduous and perennials cannot be well defined or on sites where species with deciduous behaviours present quasi-perennial behaviours in other regions. These particularities must be recorded, but it should be noted that the dominant regional condition is the key to denoting the phenological condition. Such a situation becomes even more complex where a study is carried out during the rainiest part of the year given that, in such circumstances, all species can, at a first glance, appear to be perennial. In such a situation, the seasonal condition of a climate and a previous knowledge of the species which are deciduous would help peg, without hesitation, the type of dominant phenology of the vegetation being studied. In cases of doubt, it suggested that the site should be examined in both the dry and wet seasons in order to note the dominant foliage condition.

The three previous levels, although tending to characterize the regional condition, are basically for use in formulating a detailed (local) vegetation study. Moreover, they permit a hierarchical orientation of regional works in which the objectives do not aspire to a refined study, but which can serve as a guide, a thematic layer of information on cover and vegetation at the mesoscale level (between 1;100,000 and 1:500,000).

Preparative Phase for Field Work

For Level IV onward, it is inevitable that an in-depth field study will have to be carried out, following an exhaustive method and with expert knowledge of the floristics to be studied. It is therefore crucial to highlight that a vegetation study pursues specific objectives and that it is thus impossible to follow one single valid method, as illustrated by the range of methods used in the study of vegetation (see it in Appendix A.7). That said, there come below a series of recommendations intended as a guide for vegetation baseline studies, useful for any objective and of great advantage for the process of making an inventory of vegetation types in Mexico.

Level IV (Leaf morphology) is a fundamental criterion in three senses—first, because it is the most widely used in previous proposals covering Mexico and a good part of the New World, second, because it provides perceptible characteristics at a glance and third because leaf morphology is the direct result of preponderant climatic, edaphic, and hydric conditions. In order to avoid confusion over how to denominate the leaf types, it has been decided to adopt a specialized technical language which can still be used by non-experts through following the guide described in Tables 3.1–3.8.

Other inherent and relevant characteristics in this level cover the development of succulence and the presence or absence of spines. These characteristics are noted during field work, which requires following a strategy which, in general, includes defining the sample site as well as the sample itself.

Sampling Strategy

The sampling strategy in vegetation studies has two phases, one diagnostic or descriptive and the other a synthesis or analysis. They are serial given that the second makes no sense without the first and vice versa. There are many texts describing the diversity of sample unit types, among which books there are two major outstanding unit types: those with a known area and those without specific area (see Appendix A.7). The diversity of the sample units, called the _relevés" in the language most commonly used in vegetation studies, owes its origin to the need to cover the various levels of biological organization from the levels of genetic complexity to the biotypes. Additionally, there is a plethora of questions, ranging from those of a purely academic nature to those concerning the means of reproductive strategy or dispersal or pollination mechanisms which make necessary units different from those useful in the definition of management strategies such as the stand in forest studies or the plot in agroforestry studies. A detailed review of each of the sample unit types does not fall within the scope of this work.

For the purposes of this work, however, it is of prime importance to review the sample unit or _relevé", coined as the only and official term for phytosociological studies. The sample/sampling unit is denominated _relevé" or phytosociological _summary/statement/inventory/plot and is based on the selection of a minimum area delimited by its homogeneity and ecological uniformity. According to Gordon 1968 as cited in Werger (1974), "*The relevé is a set of ecological observations concerning a determined site*"‖. The reader may wonder as to the reproduction of the same type of sample unit. There are three reasons: one is that there exist today around one million phytosociological relevés which have been studied worldwide in accordance with an established methodology; another is the simplicity of its execution, which permits its replication even in substantially different ecosystems;

and the last refers to the benefits of comparison aiding in management and conservation policy-making based on the standardized hierarchical analysis of vegetation (syntaxonomy) and a congruent territorial expression (synchorology) even across enormous territories such as that of the European Union.

Diagnostic or Descriptive Phase

The selection of the relevé site: this aspect, although it may appear simple, is very complex, given that selection depends on the objectives, scope, and goals of the study. For a vegetation baseline study which, moreover, has as one its goals a coherent cartographic expression, it is recommended to follow a series of steps:

- The delineation of coverage previous to the field work based on the interpretation of characteristics measuring the reflectance of the vegetation by means of geographic imaging (aerial photography, radar, or satellite imagery) will be of great assistance in three senses—in enabling stratification of the categories from cover to vegetation, thus making it possible to draw on a stratified sampling design; in allowing sample optimization to the point of including in the sampling strategy a greater coverage diversity and then being able to access it when not in the field; and in facilitating cartographic expression obtained from the field results based on a stratified design. The cartographic expression comprises a data based on polygons grouped into classes which, by their spectral characteristics are considered homogeneous to the interior and distinct from their context, which assists in generating a stratified sampling design a priori. On site (that is on a 1:1 scale), the complexity is not diminished, and thus, the selection of the relevé site to be carried out a posteriori should be made in line with the following criteria:
- The gathering of information on the complexity of vegetation in situ should always be carried out in light of the abstract construction of the site, where, at the same time, the heterogeneity of the context is visualized, but with homogenous spaces sought, which, in their totality, better represent said heterogeneity. In this way, research on a heterogeneous site will require a greater number of relevés to study until the complexity is fully described. In each space, one or a number of relevés will need to be made, which, based on the information they in their totality yield, will have to be carried out a number of times in order to obtain a series of _repetitions" generating a representation of the site's floristic composition, structure, and function, describing homogeneous areas included within the site. It should be noted that the term _repetition" refers to a homogeneous condition in terms of floristic composition, structure, and function.
- One of the most common problems is the selection of the relevé site in a context where the prevalent conditions of disturbance of anthropic origin are expressed in secondary vegetation stadia. If the ultimate goal is the creation of a current

vegetation map, the relevés must include each of the disturbance conditions, while, at the same time, if a reconstruction of the native vegetation distribution pattern is desired, the survey will favour the remaining preserved sites by avoiding those with prevalent, widely distributed taxa (ruderals and arvenses) which usually take advantage of disturbed areas.

- In selecting the relevé size, it is necessary to delimit a homogeneous area where it is possible to carry out a complete inventory of trees, shrubs, grasses, and flush vegetation. The form of the relevé must adjust itself to the criterion of homogeneity. In practice, many texts suggest limits of quadrants of various sizes depending on the dominant physiognomy: for forests, between 500 and 1,000 m^2; for shrubland between 100 and 500 m^2, and for grassland between 20 and 100 m^2 (Mueller-Dombois 1974).
- The sample size or number of relevés becomes a task of clinical reasoning given that it is not possible to know a priori the minimum number of relevés needed to describe the total heterogeneity, given that it is unknown. For practical effects, it is recommended to begin with no fewer than five relevés per cover category. If the floristic composition and structure between the relevés within the same cover category differ substantially (a significant replacement in the species composition or structure—for example, of a forest or shrubland), then it will be necessary to carry out more relevés in conditions where fewer data are available. The refined analysis of this information is carried out in the following phase, but, at the same time, the attempt is being made to ensure that the greatest heterogeneity of the vegetation present is included.
- The type of data to be collected in each relevé is the most critical part of this proposal, given that, in order to be comparable, it must include a minimum of information collected according to the same methods and criteria. The complete floristic inventory of each relevé is indispensible and inevitable in order to be able to carry out correctly the phase of analysis and synthesis. In the field, it will be very unlikely that a researcher would know all species types in their diverse stadia, and so it is recommended to use vernacular names for the species where the scientific name is unknown, to collect examples for determination in the herbarium as well as to construct a miniherbarium where a sample is shown with its assigned vernacular name. For non-fertile samples, it is necessary to collect reference samples until one with flowers, fruits, or seeds is found for determination.
- The second fundamental datum refers to measuring dominance which, in general, is denoted by density, cover, and frequency or a combination thereof (Value of Importance). The measure adopted must adapt itself to the objectives of the study, but the recommendation for the purposes of a baseline study is to quantify the relative coverage, which term refers to the land surface covered by a species (the set of all the individuals within a relevé). This quantification is usually expressed in percentages or even in percentage classes and is done with the aim of increasing exactitude (though precision is lost) and the pace at which data are collected. It furthermore reduces the subjectivity, given that the assumption that each species falls within a class incorporates a lower error

margin than if the attempt is made to measure the absolute cover. The relative coverage has two specific purposes: on the one hand, it is a direct measurement of the abundance–dominance of a species in a relevé (in analytical terms corresponding to the unimodal behaviour which theoretically explains the distribution of the vast majority of the species and thus is an adequate measurement of a given condition at a given time); on the other hand, it is an attribute permitting the abundance of a species to be translated into an expression of —coverage‖ (here referred to as the expression of the reflectance value derived from a photograph or image) of the community to which it is finally associated (the meaning of ‗association" in the broad sense).

- The rest of the data shown in Appendix A.8 are obligatory for any type of study and are not greatly conflictive. It should be added that, for studies attempting to identify the variables best explaining the vegetation patterns to be described, it is recommended that said direct variations (humidity, soil depth, disturbance among others) as well as indirect variables (site-use history, tenure, among others) should be estimated. Some of these variables may be measured on site; others must be gathered from other sources such as local people, literature, and existing remote-sensing images.

Analysis and Synthesis Phase

The Raw Table

The information collected directly in the field can be analysed in two ways simultaneously. The first and crucial way concerns the process of determination of the samples. This is laborious and requires both regional flora experts and taxonomists expert in specific groups and in comparing with examples from the herbarium. In regions where floristic knowledge is still limited, it is suggested to continue with collections of each of the outputs from the field, despite already knowing the taxonomic determination of the sample in question. The second analytical front is where the entry of the information collected into some form of database is made, permitting, on the one hand, identification by a unique floristic name (preferably with names of species already collected) and, on the other hand, the relative coverage values of each taxon to be registered. This sequential arrangement of each relevé (columns) with each taxon (rows) and their corresponding coverage value (each cell) is known as a —raw table‖ (sensu Mueller-Dombois 1974).

The raw table is characterized by still maintaining a number of taxa even in the determination process, the common or vernacular names thus able to be used. These common names can be substituted by formal ones as the taxonomic determination is being completed. Moreover, the raw table permits a distinction at a glance between the taxa (be they common or scientific names) which are presented with a constant frequency in a high percentage of the relevés (e.g. taxa present in more than 70 % of relevés). Equally, the taxa with a very low frequency (e.g. taxa present in fewer than five per cent of all relevés) are distinguished. High-percentage taxa are denoted as companion species or constants independently of their coverage value throughout the relevés in total; low-percentage taxa are denominated as rare species. In this way, the complexity of the database is, in general, reduced substantially (for places where the dominance of a group of species is not so obvious) to significantly (for places with very markèd dominance by some species). This process does not eliminate the determination of the entire sample set collected, but it does permit prioritization of the work given that the companion and rare taxa will not be the priority for the purposes of the analysis a posteriori.

Partial Tables

This process will be followed by a first tabular arrangement in such a way that all the rare species will be cut and pasted into the basal part of the raw table. Above, the rare species come the companion species and, in this way, the construction of semi-arranged partial tables is begun. The functional procedure is for the relevés corresponding to the same territorial or cartographic unit (e.g. of the same polygon or of the same landform or those which are found in the same type of soil) to be placed in a continuous manner. This type of tabular arrangement can be interesting and tedious given that, each time the order of the relevés is altered, a different order in the species groups will be observed. At this moment, the construction phase of partial tables must be considered to be concluded, now passing on to the following phase—working towards an arranged table.

The Arranged Table

The only objective of this part of the analytical process consists of identifying the species groups which, by their constant presence in relevés and their performance expressed in their coverage values in the same relevés, make up societies. These are denominated phytosociological groups. In its turn, there will be a unique set of relevés better agglutinated into a given phytosociological group. The totality of a phytosociological group and its relevés represent the association. As a result, the arranged table permits with greater or lesser clarity the group of associations occupying the territory studied to be distinguished. The theoretical postulate underlying the position that associations exist was described in Braun-Blanquet (1979) and later modified from the gradient perspective by Whittaker (1973). Both concur that the distribution of plant species occurs throughout conditions which change gradually and that there exist certain thresholds denoting the limits on sensitive species. The thresholds express functional processes and on the land, they are described as ecotones or transitional zones. In cold or temperate ecosystems with little or no anthropic incidence, the ecotones are reduced, and the turnover of characteristics is obvious, even without the use of analytical tools based on mathematical algorithms. This situation is reversed in the event of a study of tropical ecosystems, where ecotones form continua which make it difficult to discriminate between associations even with the assistance of robust algorithms. In published research, there abound methods both empirical and statistical which assist in the identification of the greater or lesser affinity between relevés and species (see Kent and Coker 1992 for a thorough review)—which methods are simply adaptations of the tools already existing to measure (dis)similarity between cases and variables. In a vegetation study, this analytical part is denominated _classification”. Its objective is double: to identify the association or plant communities from the phytosociological group, characterizing them and their corresponding relevés and also to measure statistically the singularity of each association with respect to the rest of the associations discriminated. The expression of this type of analysis consists of showing by means of graphs the affinity relationship maintained by the totality of the associations of the region studied. A detailed explanation of the algorithms used (Ludwig and Reynolds 1988; Magurran 2003) the software available (Kent and Coker 1992) and of the various graphs showing the results derived from this type of plant-classification analysis has led to a dozen books and are beyond the scope of this chapter. The goal is achieved when it is possible to discern between the existing associations as well as the eight levels of aggregations of the associations (series of associations or alliances).

From this discrimination, the next step is to make a detailed description of each series of associations and of each association with emphasis on the species set typifying them (characteristic, exclusive, and preferential species). This implies that said result should be expressed in a final and single tabular arrangement where the characteristic species head the top rows of the table, and the relevés follow in strict order of affinity measured by some type of analytic tool. In the orthodox phytosociological approach, a unique name is then coined for each association based on the characteristic species (syntaxonomy). It is also usual to identify the type of each association to the relevé, that is that which best represents the characteristics of the association in its totality (Table 4.4). At this moment, the analysis of classification is concluded, leading to a series of questions on possible factors to explain the distribution pattern of the associations described. This process is denominated _ordination” and, as with classification, it is based on a series of mathematics-based support tools, making it possible to (cor)relate the discriminated associations with possible causal factors for the distribution pattern.

It should be explained that classification and ordination are together known as the multivariate gradient analysis. The correct discrimination of the associations beyond the algorithm used depends exclusively on solid floristic work carried out beforehand. On the other hand, the factor(s) best explaining the distribution pattern of the discriminated associations will depend on the set of measured variables, on the type of measuring carried out and on the possible correlation between each one of the factors. These analyses are denominated _explanatory”, given that

Table 4.4 Criteria proposed by Navarro and Maldonado (2002) for the selection of characteristic or indicative species and priority sample species in the field

Criteria	Description
Ecological amplitude	Species restricted or almost restricted to determined environments or with a maximum distribution in determined ecological conditions (bioclimatic, edaphic, altitudinal etc.)
Physiognomy	Species contributing substantially to the appearance or physiognomy of the vegetation
Endemism	Species of areas endemic or subendemic to the distribution territory of the studied vegetation types
Phytosociological fidelity	Species with a high consistency of presence in a determined vegetation types
Centres of origin and biogeographic diversity	Species belonging to the group of floristic elements with its centre of origin and/or diversity in the biogeographic unit to which the vegetations types studied belong
Succession dynamic	Species belonging to the specific structural dynamic of the vegetation studied (forest, shrub, herbaceous, etc.)
Taxonomic identification	Species at least moderately easy to recognize and identify taxonomically in the field or with a sufficient level of taxonomic in the herbarium

deterministic conclusions in general are not reached through this type of procedure. One way of attaining deeper understanding is to redesign a verification and calibration strategy leading to a new phase of field work to corroborate the level of certainty as to the inferred assertions.

In studies where the ordination results succeed in identifying the variables, explaining to a large degree the distribution pattern of the associations, it is usual to think of predicting the distribution of these in territories not floristically studied but where the variables are known. Additionally, the cartographic expression of the distribution pattern of the associations and their causal factors can be incorporated. The prediction modelling for vegetation patterns as well as correct cartographic expression is broad topics, complex in terms of the tools required, and laborious when it comes to the methods used.

References

Braun-Blanquet J (1979) Fitosociología. Bases para el estudio de las comunidades vegetales. Blume, Madrid.

COTECOCA (1994) Revegetación and reforestación de las áreas ganaderas en las zonas áridas and semiáridas de México. Comisión Técnico Consultiva para la definición de los Coeficientes de Agostadero. México.

Dinerstein E, Olson D. Graham D, Webster A, Primm S, Bookbinder M, Ledec C (1995) Una evaluación del estado de las eco-regiones terrestres de América Latina and el Caribe. Banco Mundial, WWF. Washington D.C.

FGDC (Federal Geographic Data Committee) (2006) National Vegetation Classification Standard, Version 2—Working Draft. Vegetation Subcommittee. Federal Geographic Data Committee, August 2006.

FGDC (Federal Geographic Data Committee) (2008) National Vegetation Classification Standard, Version 2. Vegetation Subcommittee. Federal Geographic Data Committee.

García E (2004) Modificaciones al Sistema de Clasificación Climática de Köppen. Instituto de Geografía. Universidad Nacional Autónoma de México. Serie Libros, 6. México, D.F.

González-Medrano F (2003). Las comunidades vegetales de México. Propuesta para la unificación de la clasificación y nomenclatura de la vegetación de México. Instituto Nacional de Ecología, México, DF.

INEGI (Instituto Nacional de Estadística Geografía e Informática) (1991) Datos básicos de la Geografía de México.

Kent M, Coker P (1992) Vegetation description and Analysis. A Practical Approach. Belhaven Press, London.

Ludwig JA, Reynolds JF (1988) Statistical ecology: a primer of methods and computing. Wiley Press, New York.

Magurran AE (2003) Measuring Biological Diversity. Blackwell Publishing Limited, Malden.

Miranda F, Hernández XE (1963) Los tipos de vegetación de México and su clasificación. Bol. Soc. Bot. Méx. 28: 29–179.

Mueller-Dombois D. Ellenberg H (1974) Aims and Methods of Vegetation Ecology. John Wiley & Sons, Manchester.

Navarro G., Maldonado M (2002) Geografía Ecológica de Bolivia. Vegetación and Ambientes Acuáticos. Centro de Ecología Simón I. Patiño, Cochabamba.

Palacios-Prieto JL, Bocco G, Velázquez A, Mas JF, Takaki-Takaki F, Victoria A, Luna-González L, Gómez-Rodríguez G, López-García M, Palma M, Trejo-Vázquez I, A Peralta H, Prado-Molina J, Rodríguez-Aguilar A, Mayorga-Saucedo R, F González M (2000) La condición actual de los recursos forestales en México: resultados del Inventario Forestal Nacional 2000. Investigaciones Geográficas, Boletín del Instituto de Geografía. UNAM. 43: 183–203.

Ramamoorthy TP, Bye R, Lot A (1998) Diversidad biológica de México: orígenes and distribución. Instituto de Ecología. Universidad Nacional Autónoma de México.

Rzedowski J (1978) Vegetación de México. Limusa. México, D.F.

UNESCO (United Nations Educational, Scientific and Cultural Organization). 1973. International classification and mapping of vegetation. Paris. http://unesdoc.unesco.org/images/0000/000050/005032MB.pdf.

Werger A (1974) On concepts and techniques applied in the Zurich-Montpellier method of vegetation survey. Bothalia, 11: 309-323.

Whittaker H (1973) Ordination and Classification of Communities. Handbook of Vegetation Science 5. Junk, The Hague, The Netherlands.

Chapter 5
SECLAVEMEX Aimed at Integrating Land-cover and Vegetation Mapping

Introduction

Vegetation, either natural or cultural, from the local to global scale, portrays the baseline for land-use planning. In order to enhance the role of vegetation in land-use planning, distribution patterns in map formats need to be constructed. A vegetation map, furthermore, provides critical information on ecological, economic and cultural attributes, which, in turn, form the basis on which decision-making on environmental policies should be made (Matteucci and Colma 1982; FGDC 2005). The quality of the vegetation map ought, therefore, to be assessed. According to Küchler (1951), *The demand for vegetation maps is becoming increasingly emphatic. As a result, a growing number of such maps are becoming available, but their quality varies within wide limits.* It is widely accepted that vegetation-map quality refers to spatial, temporal, and content resolutions. In order to meet acceptable standards in vegetation mapping, a number of rules must be clearly followed, namely: hierarchical nomenclature, scale and cartographic metadata. These rules are often not followed carefully or overlooked altogether, even though the outcomes affect environmental policy choices directly. In order to avoid misleading decisions, a number of steps are recommended. This is especially critical now that a —land-cover map‖ is often considered as a surrogate for a —vegetation map‖. Technically, both depend on plants as input for mapping purposes, yet they differ significantly in spatial (scale), temporal (time window), and content (vegetation knowledge) resolutions (Pedrotti 2004, 2013).

The objective of this chapter is threefold. First, it provides a thorough review of three different but complementary concepts, namely: vegetation, land cover, and land use; second, it further provides the basis of their complementarity, and third, it uses the present SECLAVEMEX framework to illustrate the construction of a map legend which clusters land-cover and vegetation categories into a hierarchical procedure. A comprehensive land-cover/vegetation map of Mexico is included as an example of the applicability.

A. Velázquez et al., *Standardized Hierarchical Vegetation Classification*, Geobotany Studies, DOI 10.1007/978-3-319-41222-1_5

Conceptual Framework

Vegetation

Vegetation is defined as clusters of species with a specific composition and structure occurring at a given place and time. Vegetation distributes along gradients (Whittaker 1978; Oksanen and Minchin 2002), but discrete groups are distinguished based upon composition and structure (Ellenberg and Muller-Dombois 1974). Vegetation can be split into cultural and natural, the former comprising induced, selected, and domesticated species by management actions for specific purposes, whereas the latter contains species adapted to local physical conditions throughout the course of evolution. Cultural vegetation forms discrete patterns, which portray land-use productive systems such as forestry, agriculture, and grazing, whereas natural vegetation forms continuum species assemblages with fuzzy boundaries (Braun-Blanquet et al. 1947).

Cultural and natural vegetation are neither static nor stable (Velázquez et al. 2000). Transitions between cultural and natural vegetation patterns are the rule rather than the exception. Natural vegetation is rendered cultural by means of man-made actions, so-called perturbation. Cultural vegetation remains as long as perturbation is continuous in intensity and frequency. Cultural vegetation, however, may become natural if perturbation ceases either in intensity or frequency. Perturbed natural vegetation goes through a number of stages before arriving at a somewhat similar natural condition. These natural–cultural–natural vegetation transition processes form whimsical complex vegetation patterns often referred to as secondary vegetation types or secondarization (Petit et al. 2001).

Stage changes in natural vegetation are also triggered by natural events, so-called disturbance actions (e.g. hurricanes, volcanic eruptions, herbivores, and fire). These processes are known as successional development stages. The type of response to such disturbance depends upon the intensity and frequency of the events (Van der Maarel 1989).

Perturbation- and disturbance-triggering actions are expressed in patches of intermingled vegetation types occurring simultaneously. To distinguish secondarization from successional processes, an assessment (scale 1:1) is carried out. Splitting these processes from smaller scales becomes a burden often expressed into fuzzy mosaics. Maps can, therefore, end up overemphasizing either natural or cultural conditions or their transitions. In consequence, maps produced for botanical, ecological, forestry, or geographic purposes turn out to be substantially different one from the other. New remote sensing techniques and geographic information systems have played a key role in reducing dissimilarities among vegetation maps. Even so, plant-species assemblages are produced solely through the use of vegetation maps. Therefore, hierarchical classification systems are crucial in order to map individuals, populations, and communities logically. An acceptable vegetation map inevitably demands a large amount of time and abundant botanical knowledge. At a certain aggregation level, vegetation is expressed through vegetation mosaics.

Land Cover

Land cover is defined as the perception of the aerial parts of plants blanketing a portion of the Earth's surface (UNEP/FAO 1994). Remote sensing is the discipline aimed at generating databases expressing land-cover patterns (Chuvieco 1998; Lillesand et al. 2014). The perception of the plant parts visible from the air depends fully on two factors: the capability of the sensor carried by an airplane, satellite, or other such apparatus to monitor the fundamental attributes of plant life (e.g. water, chlorophyll, and temperature levels) and on the ability of the interpreter to distinguish (correlate) plant attributes—detected by the sensor—as expressions of plant species (Kent and Coker 1992). Since sensors—regardless of their spatial, temporal, and spectral resolution—are not able to distinguish attributes which depict plant species, land cover often portrays vegetation mosaics.

Vegetation mosaics are the unique objects of a land-cover map, and their attributes are portrayed by texture, pattern, and colour. These attributes, combined with the experience of the interpreter, are crucial in compiling a sound land-cover map. There are currently innumerable data available (in digital format) from sensors with different spatial, temporal, and spectral resolutions (Table 5.1). Land-cover map generation has, accordingly, increased exponentially in recent decades. The resulting maps are available in local, regional, and global scales and have been designed for different purposes, particularly those requiring a high temporal and spectral resolution. The vast majorities of these land-cover maps are constructed under laboratory conditions and are therefore not subjected to ground verification. In response, (un)certainty assessment of land-cover maps has been established as a new scientific standard in order to increase reliability. Land-cover maps are thus models with different levels of precision (degree of statistical heterogeneity reached in the analytical processes) and accuracy (measurement of exactitude in the input's representation of the real object). The general underlying rule is as follows: the more local or micro the scale, the more accuracy and the less land coverage and precision. Conversely, the more global or mega the scale, the more land coverage and precision and the less accuracy.

A thorough review of the advances in remote sensing as a major source of input for land-cover maps is far beyond the scope of this contribution. Nonetheless, it is quite certain that land-cover and vegetation maps are nowadays approached simultaneously, and the construction of the latter is fully dependent on the former.

Because of the nature of vegetation mosaics, secondarization processes are the only ones able to be delineated in land-cover maps. An acceptable land-cover map invariably demands high-quality images and in-depth knowledge of remote sensing. At certain aggregation levels, individual types of land cover are expressed as land-cover mosaics.

Land Use

According to Vitousek et al. (1997), there is no place on Earth without evidence of human activity. It is therefore undeniable that all vegetation and land-cover types

Table 5.1 Comparison of the most-used remote sensing inputs to generate land-cover databases

	Sensor	ASTER	LANDSAT-8	Qucikbird-2	SPOT-6/7	Sentinel-2
Wavelength (μm)/ band	VNIR	0.52–0.86/3	0.43–0.89/4 0.5–0.68 (Pan)/1	0.45–0.90/4 0.45–0.90 (Pan)/1	0.45–0.745 (Pan)/1 0.45–0.89 (VNIR)/3	0.443–0.94/10
	SWIR	1.60–2.43/6	1.56–2.30/3	–	–	1.375–2.19/3
	TIR	8.125–11.65/5	–	–	–	–
	VNIR	15	30 (Pan) 15	2.4 (Pan) 0.61	6 Pan (1.5)	10/4 20/4 60/ 2
Spatial resolution (m)	SWIR	30	30	–	–	20/2 60/1
	TIR	90	–	–	–	–
Radiometric resolution (bits)		8 VNIR 8 SWIR 12 TIR	12	11	12	12
Temporal resolution (Recurrent day)		16	16		1–5	10/5
Swath (km)/scene size		60	185	16.5	60	100

are, with varying degrees of intensity and frequency, subjected to man-made management actions.

According to Meyer and Turner (1996), land use is defined as concerted man-made actions to turn natural vegetation into resources. These anthropogenic processes are expressed as cultural mosaics over the Earth´s surface and vary significantly from place to place (Velázquez et al. 2010). These cultural mosaics are often highly correlated with land-cover mosaics, hence, often being named after land-use types. The more conventional ones are human settlements, cropland, forestland, and pastureland, among others.

Because of the nature of land-cover mosaics, secondarization processes are the only ones able to be delineated in land-use maps. To be of use, a land-use map requires high-quality images and interpretation by an expert in remote sensing and production systems. At certain aggregation levels, vegetation, land-cover, and land-use types are mapped into mosaics. The dominant types are chiefly used to label the polygon containing the intermingling mosaics. The label of the polygon is further compared with other polygons, and, eventually, a number of polygons comprising similar mosaics are named after the prevailing type. This cluster of polygons is recognized as the cartographic class in the legend. The process of depicting and naming cartographic classes is thus an oversimplification of the vegetation, land-cover, and land-use types contained within each class.

To summarize briefly, a land-use cartographic class comprises in a number of land-cover classes. In turn, a land-cover cartographic class chiefly comprises a number of vegetation types. This is why vegetation, land-cover, and land-use maps are significantly different, even though they depend on plants as cartographic objects.

Vegetation and Land-Cover Maps

A vegetation map, *sensu stricto*, is understood as the two-dimensional representation of individual, populations, and communities of plant species clustered into assemblages occurring at a given place and time (Lewis 1998). Academically, a map must meet all cartographic standards, such as scale, orientation, datum, projection, coordinates, and a self-explanatory legend. According to Küchler (1967), most scientific literature of the day apparently comprising a vegetation map in fact contained figures of land-cover types. Vegetation-map legends are crucial and are aim-oriented. A proper legend must rely on a hierarchical vegetation classificatory system (Bohn et al. 2000).

There are currently two approaches to constructing a vegetation map: botanical and geomatic (Wyatt 2000; Pedrotti 2004). Both assume conceptual and technical knowledge as well as the skill to display properly the vegetation complexity regardless of the study area. A combination of both approaches is highly recommended given that interdisciplinary frameworks are preferable in the production of a sound vegetation map.

The botanical approach (*sensu* Knapp 2012) seeks to achieve the cartographic portrayal of vegetation communities of similar biogeographical affinities along biophysical gradients. Its fundaments rely on a phytosociological classification framework, such that a hierarchical system applies, and it is consistent with the scale of the map (Pedrotti 2004). A small-scale map thus attempts to include rough phytosociological levels (class or subclass), whereas a large-scale map aims for detailed phytosociological levels (alliance or association). In addition, phytosociological-level distribution patterns are portrayed through factors triggered by climate, water, soils, and human activity. In consequence, climatic, hydro, edaphic, and anthropic vegetation series are outlined (Pedrotti 2004). As a result, three main types of vegetation maps are constructed: botanical (floristic), vegetation, and phytogeographical (Table 5.2).

The geomatic approach is based on the progress in remote sensing, which, over the last four decades, has revolutionized vegetation cartography. Currently, aerial photographs are rarely used, since satellite sources are readily available with a wide range of spatial, temporal, and spectral resolutions (Table 5.1). The core difference concerns polygon delineation attributes. Texture, colour, and pattern were the critical attributes in aerial photographs, whereas, with satellites, the spectral values of images were the unique attributes used to distinguish one polygon from its neighbouring polygons. In addition, spectral values are organized in pixels and pixel aggregates, forming a

Table 5.2 Hierarchical relationship between phytosociological levels and vegetation maps as recommended by the botanical approach (modified after Pedrotti 2013)

Level	Object represented	Approach/type of map	
Individual plants	Species	Autoecological	1. Botanical or floristic
Plant populations	Species	Sinecological	
Plant-species assemblages	Association (communities)	Phytosociological	Vegetation
Plant community assemblages	Alliance	Phytosociological	
Catenal	Association series	Geobotanical	
Phytogeographic unit	Plant realm	Phytogeographical	Phytogeographical
Bioma	Vegetation life zone	Phytoclimatic	

database in raster format which enables automated database storage, handling, and manipulation via a geographic information system. In consequence, the production rate for maps with proper projection and scales is significantly faster than the traditional aerial photograph procedure (Kraak and Ormeling 2003). Geomatic production might disregard the need for field data to construct the legend, and assumptions are taken where spectral responses are correlated to plant-species assemblages. This assumption is certainly inaccurate, since microscopic features distinguishing plant species such as flower components do not play a crucial role in spectral resolution. In consequence, the geomatic approach lacks a proper vegetation classification framework (Alexander and Millington 2000). This is why the geomatic approach is meant to produce land-cover rather than vegetation maps.

Discussion

Eastern and Western Europe (Pedrotti 2004) and a few other countries have systematically produced vegetation maps following a botanical approach. This has been the result of systematic vegetation sampling for a century following a phytosociological framework (Bohn et al. 2000). Outside Europe, outstanding efforts have been made by Venezuela (Hueck 1960; Hubber et al. 1984); Cuba (Borhidi 1987); and South Africa (Mucina and Rutherford 2006). Other countries, such as Bolivia (Navarro 2003), Colombia (Rangel and Minorta 2014); and Mexico (Miranda and Hernández-X 1963; Rzedowski 1978) are remarkably rich in botanical knowledge, yet this knowledge has played a limited role in rigorous botanical cartography. North America (Canada and the USA) has recently launched an effort to produce land-cover/vegetation maps following a hierarchical approach (Jennings et al. 2009).

Currently, land-cover maps produced following the geomatic approach prevail worldwide (Lapola et al. 2008). Brazil stands out for its extensive production of land-cover maps (http://www.inpe.br/), although most of them lack botanical content. Most countries are following the geomatic approach to produce land-cover maps as surrogates for vegetation maps. As previously explained, the botanical and geomatic approaches are complementary, yet significantly different in content and therefore quality of information. The outreach, especially in supporting environmental policy formulation, is limited if decisions are based purely on land-cover maps.

Efforts to map Mexico's vegetation have been prolific. By the end of the XIXth century, systematic efforts began (Ramírez 1899 quoted by Quintero 2001). Chronologically, a number of additional regional efforts were envisaged by Fournier (1900), Harshberger (1911), Ochoterena (1918), and Sanders (1921) (all these four quoted by González-Medrano 2003). Ochoterena (1937) provided an outstanding contribution map of plant distribution patterns. In chronological order, other efforts followed: Dice (1943), Smith and Johnston (1945), Goldman and Moore (1946), Leopold (1950), Arellano (1952), and Aubreville (1962). All based their cartographic efforts on physiognomic attributes at a large-scale level, since botanical data were still being gathered, and the lack of a systematic vegetation classification system became evident.

The contribution of Miranda y Hernández-X (1963) became a bench mark, since, for the first time, a thorough description of physiognomic and floristic attributes was compiled to describe the vegetation types of Mexico. This contribution served as a model for the construction of the first systematic cartographic vegetation efforts. In the 1970s, the Mexican mapping agency (INEGI, by its Spanish acronym) started a vegetation and land-use mapping program at semi-detailed and detailed scales. The botanical approach prevailed in Mexico before the 1990s, producing databases covering the classification and mapping of vegetation for the whole country. The classification schemes of Miranda and Hernández-X (1963), Rzedowski (1978), and González-Medrano 2003 are clear evidence of the botanical approach. After the 1990s, as remote sensing technology progressed, the geomatic approach began to prevail, pioneered by Velázquez and Cleef (1993) and Mas et al. (1996); more recently, there have come into existence Mexico-wide semi-automated geomatic databases, which have begun to become readily available for example, the ones produced by Mas et al. (2004), Sánchez-Colon et al. (2009), and Velázquez et al. (2010). None of these efforts, nonetheless, surpasses the systematic effort conducted by INEGI (2005) as described in the five vegetation series at scale 1:250,000 so far compiled.

Outreach and Implications

Classification and vegetation-map legends are closely related, yet they are significantly different. The former is based on botanical attributes scale 1:1, whereas the latter is vegetation map-scale-remote sensing source dependent. Nowadays, both are complementary, and a rigorous vegetation map is unthinkable without its classification counterpart. The increasing demand for vegetation maps extends to disciplinary fields such as forestry, agronomy, soil science, geology, ecology, and climatology among other. In the socio-economical realm, land-use planners, investors, economists, politicians, and the military are some of the most demanding fields seeking sound vegetation maps as a basis upon which to make critical decisions (Pedrotti 2004). This is especially true for the most critical global issue, understanding, monitoring, and predicting the effects of climatic change.

At this point, the SECLAVEMEX framework's usefulness in facilitating the transition by way of a formal classification scheme to a vegetation-map legend becomes evident. The first three levels (Bioma, Mayor Formation, and Formation) are meant to be easily delineated by means of remote sensing and climatological data sources. At these three levels, cartographic scales are macroregional (between 1: one to four millions scales), covering surfaces of the order millions of hectares. Field work is optional if available databases are reliable for accuracy validation and in consequence increasing certainty of the land-cover classification. This is the case here illustrated in the three maps attached as appendices seamlessly unifying the geomatic and the botanical approaches. These three maps may be visualized in detail and printed at a scale of 1:4,000,000 (http://www.ciga.unam.mx/landcover/mapping/). The hierarchically organized legend is the logical result of attributes describing the uniqueness of each vegetation type.

Levels fourth and fifth (subformation and series of associations) aimed at attending mesoregional scales (between 1: one hundred thousand to half a million scales), covering surfaces of the order to thousands of hectares. At this scale, field work must be carried out to verify dominant vegetation morphology, presence/absence of spines, and succulence of dominant taxa and eventually to depict prevailing families or genera in each cartographic class.

Levels sixth, seventh, and eight (association, subassociation, and facie) aimed at taking care of local microregional scales (between 1: ten thousand to fifty thousand scales), covering surfaces of the order to hundreds of hectares. At this scale, phytosociological relevés are strongly recommended as strategy to conduct field work. The target focuses on depicting unique combination of species assemblages. Cartographic classes—depending on the degree of heterogeneity and disturbance—will comprise vegetation mosaics.

Western Europe is so far the unique continent where vegetation mapping is likely to occur at all eight levels (Gaussen 1948, 1953). Other large regions such as North America are compiling databases to obtain similar detailed vegetation maps. The hierarchy and attributes of the SECLAVEMEX here described are meant to serve as guideline to organize databases and accordingly produce meaningful vegetation maps. Meso- and microregional examples of vegetation maps are abundant in Mexico as the case in many other countries, yet comparison among them is doubtful since criteria are incomparable (FAO 1995). It is therefore the desire of the authors that the SECLAVEMEX framework bridges logically geomatic and geobotanical vegetation mapping approaches. Conceptually, we may build a comparable base line worldwide to precisely assess current trends on native vegetation loss. Eventually, we may also target at places where disturbance and perturbation are critical. All this is fundamental to predict temporal and spatial dynamics of land-cover/land-use and vegetation types as response to the indisputable climatic changes foreseen.

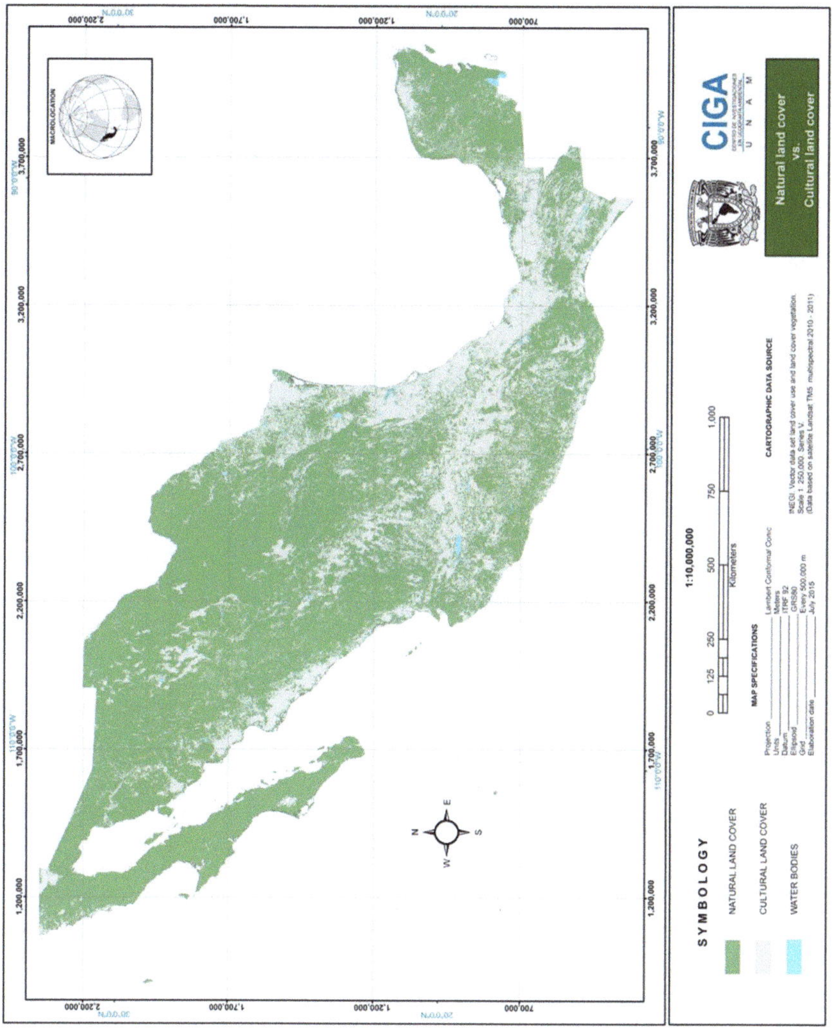
MEXICO
MACROLOCATION
SYMBOLOGY
NATURAL LAND COVER
CULTURAL LAND COVER
WATER BODIES
1:10,000,000
0 125 250 500 750 1,000
Kilometers
MAP SPECIFICATIONS
CARTOGRAPHIC DATA SOURCE
CIGA
Natural land cover
vs.
Cultural land cover

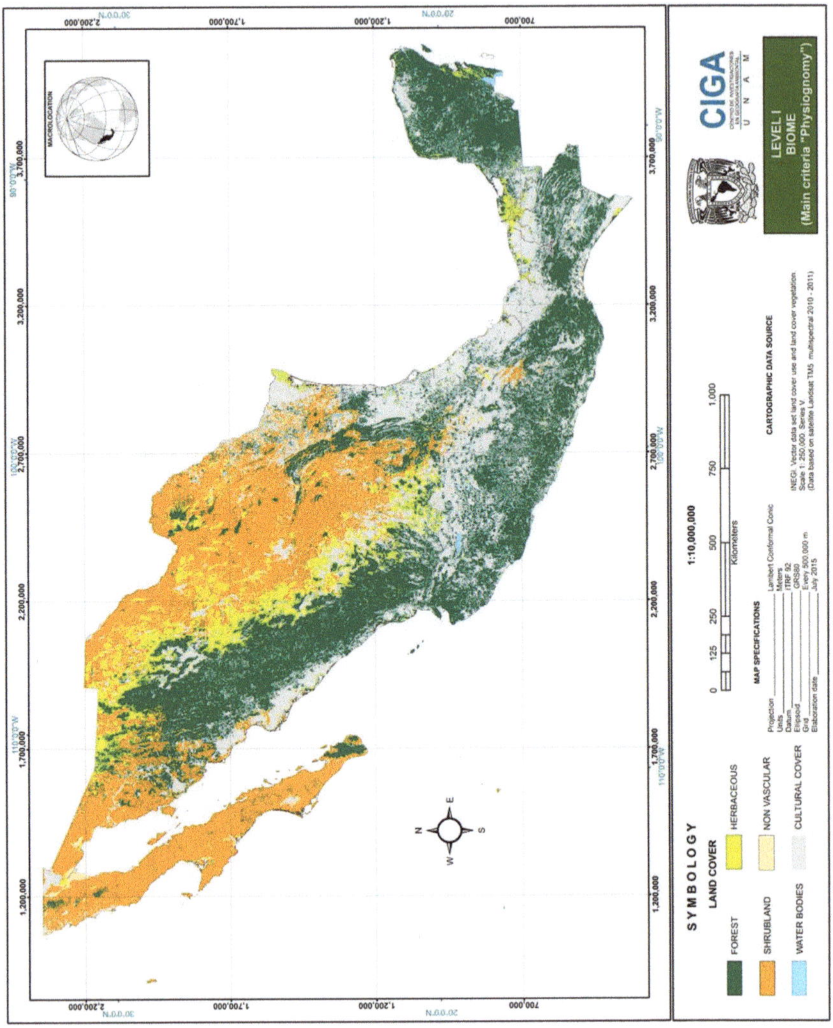
MEXICO
MACROLOCATION
SYMBOLOGY
LAND COVER
FOREST
SHRUBLAND
WATER BODIES
HERBACEOUS
NON VASCULAR
CULTURAL COVER
1:10,000,000
0 125 250 500 750 1,000
Kilometers
MAP SPECIFICATIONS
Projection Lambert Conformal Conic
Units Meters
Datum ITRF 92
Ellipsoid GRS80
Grid Every 500 000 m
Elaboration date July 2015
CARTOGRAPHIC DATA SOURCE
INEGI. Vector data set land cover use and land cover vegetation
Scale 1: 250,000. Series V
(Data based on satellite Landsat TM5 multispectral 2010 - 2011)
CIGA
U N A M
LEVEL I
BIOME
(Main criteria "Physiognomy")
N
E
S
W
1,200,000
1,700,000
2,200,000
2,700,000
3,200,000
3,700,000
700,000
110°0'0"W
100°0'0"W
90°0'0"W
20°0'0"N
30°0'0"N

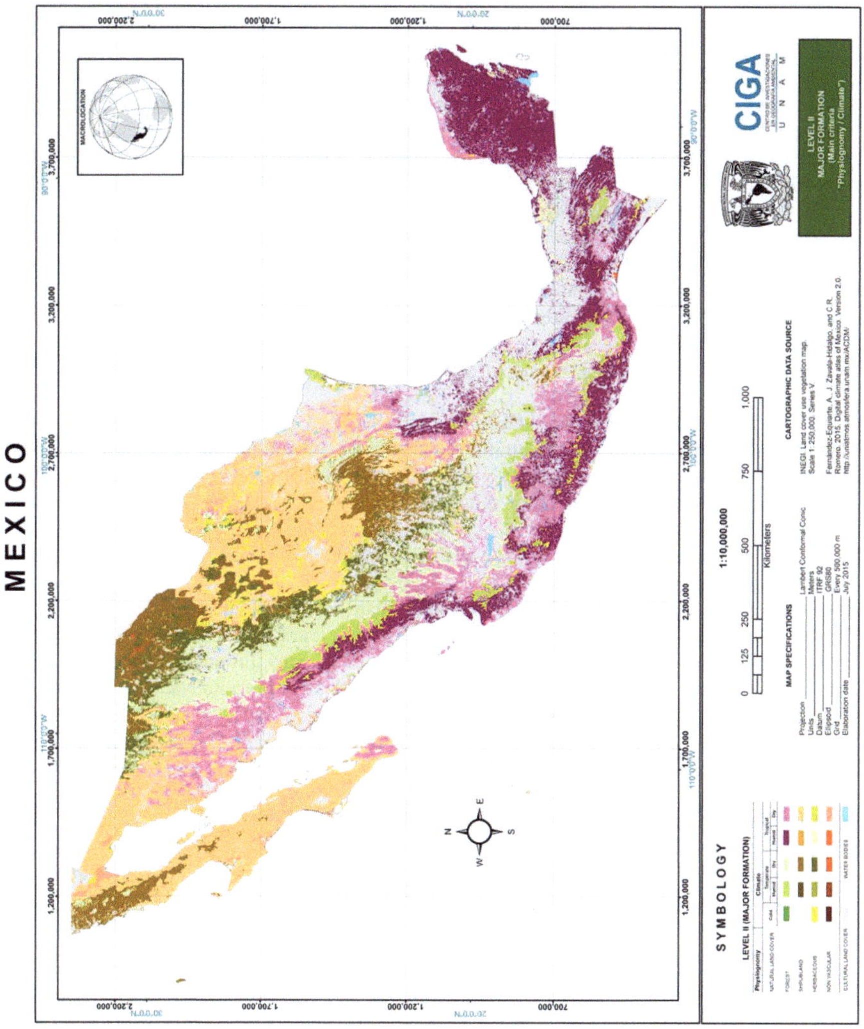
MEXICO
SYMBOLOGY
LEVEL II (MAJOR FORMATION)
1:10,000,000
Kilometers
MAP SPECIFICATIONS
CARTOGRAPHIC DATA SOURCE
CIGA
LEVEL II
MAJOR FORMATION
(Main criteria
"Physiognomy / Climate")

References

Alexander R, Millington AC (eds) 2000 Vegetation Mapping. John Wiley & Sons, Manchester.
Arellano L.Macías (1952) The Future of Mexico Lies in Education, In: Transactions of the Fifteenth North American Wildlife Conference, 9–10.
Aubreville A (1962) Temas fitogeográficos: 1. Clasificación fisonómica and florística de las formaciones vegetales tropicales: 2. Clasificación ecológica de las formaciones vegetales tropicales, 3. Clasificación de las principales formaciones vegetales de México. Inst. Mex. Rec. Nat. Renov. A.C. México, D. F.
Bohn U, Gollub G, Hettwer C, Neuhäuslová Z, Raus T, Schlüter H, Weber H (2000) Karte der natürlichen Vegetation Europas, Maßstab 1: 2 500 000. [Map of the Natural vegetation of Europe. Scale 1: 2 500 000]. Bundesamt für Naturschutz, Bonn.
Borhidi A (1987) The main vegetation units of Cuba. Acta Botanica Hungarica 33(3), 151–185.
Braun-Blanquet J, Emberger L, Molinier R (1947) Instructions pour l'établissement de la carte des groupements végétaux. Centre Nacional de la Recherche Scientifique, Paris.
Chuvieco E (1998) El factor temporal en teledetección: evolución fenomenológica y análisis de cambios. Revista de Teledetección, 10: 1–9.
Dice IR (1943) The Biotic Provinces of North America. Ann Arbor. Michigan.
Ellenberg D, Mueller-Dombois D (1974) Aims and methods of vegetation ecology. Wiley Emmart, New York.
FAO (1995) Planning for sustainable use of land resources. Towards a new approach. FAO Land and Water Bulletin, 2.
FGDC (2005) Nacional Vegetation Classiffication Standard Hierarchy Revisions Woorking Group. Federal Geographic Data Committee, Vegetation Subcommittee. OverView July 2005.
Fournier EPN (1900) Mexicanas Plantas Nuper a Collectoribus Expeditionis Scientificae Allatas aut Longis ab Annis in Herbario Musei Parisiensis Depositas Praeside J. Decaisne... Enumerandas Curavit. Parisiis : ex Typographeo Reipublicae, 1872–1886.
Gaussen H (1948) Carte de la végétation de la France, Feuille Perpignan. Service de la carte de la végétation de la France, Tolouse.
Gaussen H (1953) A proposed ecological vegetation map. Surveying and Mapping 13: 168–173.
Goldman EA, Moore RT (1946) Biotic provinces of Mexico. Journal of Mammalogy 26: 347–360.
González-Medrano F (2003) Las comunidades vegetales de México. Propuesta para la unificación de la clasificación y nomenclatura de la vegetación de México. Instituto Nacional de Ecología, Secretaría de Medio Ambiente y Recursos Naturales. México, D.F.
Harshberger JW (1911) Phytogeographic Survey of North America. G. E. Stechert, New York.
Huber O, Steyermark JA, Prance GT, Alès C (1984) The vegetation of the Sierra Parima, Venezuela-Brazil: some results of recent exploration. Brittonia, 36(2): 104–139.
Hueck K (1960) Vegetation map of Venezuela. Bol. Inst. For. Latino Amer. Merida 7: 3–15.
INEGI (Insitituto Nacional de Estadística Geografía e Informática). (2005) Estados Unidos Mexicanos. Principales Avances en Materia Cartografica. VIII Conferencia Cartografica Regional de las Naciones Unidas para America, New York.
Jennings MD, Faber-Langendoen D, Loucks OL, Peet RK, Roberts D (2009) Standards for associations and alliances of the US National Vegetation Classification. Ecological Monographs 79(2): 173–199.
Kent M, Coker P (1992) Vegetation description and Analysis. A Practical Approach. Belhaven Press, London.
Knapp PR (2012) II. Vegetation Science (Sociological Geobotany). Progress in Botany/Fortschritte der Botanik: Morphology Physiology Genetics Taxonomy Geobotany/ Morphologie Physiologie Genetik Systematik Geobotanik 44: 418.
Kraak MJ, Ormeling F (2003) Cartography: Visualization of Geospatial Data. Prentice Hall, Glasgow.
Küchler AW (1951) The relation between classifying and mapping vegetation. Ecology 32: 275–283.
Küchler AW (1967) Vegetation Mapping. Ronald Press Company, New York.

Lapola DM, Oyama MD, Nobre CA, Sampaio G (2008) A new world natural vegetation map for global change studies. Anais da Academia Brasileira de Ciências 80(2): 397–408.
Leopold AS (1950) Vegetation zones of Mexico. Ecology. 31: 507–518 (1735–1770).
Lewis MM (1998) Numeric classification as an aid to spectral mapping of vegetation communities. Plant Ecology 136: 133–149.
Lillesand T, Kiefer RW, Chipman J (2014) Remote sensing and image interpretation. John Wiley & Sons, Manchester.
Mas JF, Sorani V, Alvarez R. (1996) Elaboraciòn de un modelo de simulaciòn del proceso de deforestacio. Investigaciones Geográficas (Mx), octubre, especial 5 Universidad Nacional Autónoma de México Distrito Federal, México.
Mas JF, Velázquez A, Díaz-Gallegos JR, Mayorga-Saucedo R, Alcántara C, Bocco G & Pérez-Vega A (2004) Assessing land use/cover changes: a nationwide multidate spatial database for Mexico. International Journal of Applied Earth Observation and Geoinformation 5 (4), 249–261.
Matteucci SD, Colma A (1982) Metodología para el estudio de la vegetación. Serie Biología OEA. Monografía 22. Washington.
Meyer WB, Turner II BL (1996) Land-use/land-cover change: challenges for geographers. Geo Journal 39(3): 237–240.
Miranda F, Hernández XE (1963) Los tipos de vegetación de México and su clasificación. Bol. Soc. Bot. Méx. 28: 29–179.
Mucina L, Rutherford MC (2006) The vegetation of South Africa, Lesotho and Swaziland. South African National Biodiversity Institute.
Navarro G (2003) Tipología fluvial y vegetación riparia amazónica en el departamento de Pando (Bolivia). Revista Boliviana de Ecología y Conservación Ambiental, 13: 3-29.
Ochoterena I (1918) Las regiones geográfico-botánicas de México. Bol. Soc. Geogr. Estd. 8: 221-231.
Ochoterena I (1937) Esquemas biotípicos and Sinecias características de las regiones geográfico-botánicas de México. An. Inst. Biol. Mex. 8: 463–597.
Oksanen J, Minchin PR (2002) Continuum theory revisited: what shape are species responses along ecological gradients? Ecological Modelling 157(2), 119–129.
Pedrotti F (2004) Cartografia Geobotánica. Pitagora, Bologna.
Pedrotti F (2013) Plant and vegetation mapping. Springer, Heidelberg.
Petit C, Scudder T, Lambin E (2001) Quantifying processes of land-cover change by remote sensing: resettlement and rapid land-cover changes in south-eastern Zambia. International Journal of Remote Sensing, 22(17): 3435–3456.
Quintero JR (2001) Apuntes sobre la evolución de la cartografía en México. Notas (INEGI) 13: 41-G.
Ramírez J (1899) La vegetación de México. Secretaría de Fomento. México, D.F.
Rangel-Ch O, Minorta-C V (2014) Los tipos de vegetación de la Orinoquia colombiana. En: J.O. Rangel-Ch. (ed.). Colombia Diversidad Biótica XIV. La región de la Orinoquia de Colombia: 533-612. Universidad Nacional de Colombia – Instituto de Ciencias Naturales. Bogotá D.C.
Rzedowski J (1978) Vegetación de México. Limusa. México, D.F.
Sánchez Colón S. Martínez AF, Leyva IC, Velázquez A (2009) Estado y transformación de los ecosistemas terrestres por causas humanas. Capital natural de México, 2: 75–129.
Sanders EM (1921) The natural regions of México. Geog. Rev. 11: 212–226.
Smith AC, Johnston IM (1945) A phytogeografic sketch of Latin America. Plants and plant science in Latin America, 11–18.
UNEP/FAO (1994) Report of the UNEP/FAO Expert Meeting on Harmonizing Land Cover and Land Use Classifications. Geneva, 23–25 November 1993. GEMS Report Series, 25.
Van der Maarel E (1989) Vegetation dynamics: patterns in time and space. In: Temporal and Spatial Patterns of Vegetation Dynamics. Springer: 7–19.
Velázquez A, Cleef AM (1993) The plant communities of the volcanoes Tláloc and Pelado, México. Phytocoenología 22: 145–192.

Velázquez A, Giménez de Azcárate J, Weinmann ME, Bocco G and van der Maarel, E. (2000) Vegetation dynamics on Paricutin, a recent Mexican volcano. Acta Phytogeographica Suecica 85: 71–78.

Velázquez A, Mas JF, Bocco G, Palacio-Prieto JL (2010) Mapping Land Cover Changes in Mexico, 1976–2000 and applications for guiding environmental management policy. Singapore Journal of Tropical Geography 31: 152–162.

Vitousek PM, Mooney HA, Lubchenco J, Melill JM (1997) Human domination of Earth's ecosystems. Science 277(5325): 494–499.

Whittaker H (1978) Classification of plant communities. W. Junk The Hague. The Netherlands.

Wyatt K (2000) Vegetation mapping from ground, air and space - Competitive or complementary techniques? In: Alexander, R. & Millington, A.C. (eds). Vegetation Mapping, John Wiley & Sons, Chichester.

Epilogue

The tendency toward environmental deterioration, currently seen as the most severe in the history of humanity, coupled with climate-change processes is illustrated through high rates of loss in terms of natural capital (Mayr 2000). In their turn, increased migration and disparities in energy consumption among different nations result in high rates of cultural capital loss (Ehrlich and Ehrlich 2005). The depletion of natural and cultural capital leads inexorably to reflection on new ways of understanding the cultural–natural dynamic and, eventually, to the search for alternatives to reverse the trend of anthropogenic degradation.

Environmental deterioration is a global issue with enormous local repercussions. On the one hand, the most industrially developed nations are identified as the causative agents behind the most major changes, while, on the other, developing nations are defined as industrial-impact buffer zones (Velázquez and Picón 2010b). Mexico, a country transitioning to industrialization, is undergoing a process of accelerated environmental deterioration, as seen through the loss of native natural capital and hence the loss of buffering capacity (Velázquez et al. 2010a). The challenge of being able to discern precision what is being lost is not trivial, given that there is no consensus on how to define organizational levels for natural capital. The framework underlying the theme developed in this book, —SECLAVEMEX‖, is an attempt to remedy this lack of definition. A rigorous academic implementation would affect various issues laid out below.

Vegetation types, as much in Mexico as in a large part of the world, are considered as the best indicators of ecosystem diversity within a given territory. Even so, there exists no global classification system sustained by rigorous definitions and typologies for ecosystems. For countries with low alpha and beta diversity, despite their size (e.g. Canada, Russia, Argentina, Australia, and among many others), the lack of a classification system is irrelevant, given the limited range of vegetation types discernible through remote sensing and the relatively few species distributed across enormous spaces. The task of discerning vegetation types in countries with high alpha and beta diversity (e.g. Madagascar, Colombia, Mexico, and among many others) is highly complex. There are many reasons, but three are irrefutable, namely the wide-ranging diversity of endemic species, the high rate of species replacement between contiguous sites, and the gradual substitution of vegetation

A. Velázquez et al., *Standardized Hierarchical Vegetation Classification*,
Geobotany Studies, DOI 10.1007/978-3-319-41222-1

types along gradients (vegetation series), which impedes accurate discernment by remote sensing devices.

The aim of developing a single, accurate inventory of the natural capital in a given territory necessitates a rigorous systematization of vegetation-related information. In this contribution, the case of Mexico is presented, given that the previously proposed systems could not rely on the quantity of information now readily available. The necessity of cartographical expression was irrelevant, and the nomenclature was limited in terms of global context (Jardel et al. 2002; González-Elizondo et al. 2007).

—SECLAVEMEX‖ contributes to the definition of levels of vegetation organization from two overall viewpoints. The first is the geographic vision beginning with the definition of major biomes based on physiognomic criteria discernible with a high degree of certainty via remote sensing images; the second is the botanical vision stemming from profound knowledge of species groupings in given territories denoting various levels of groupings along catenae. This —geobotanical‖ approach was lacking in the education of biological diversity experts in Mexico and is a key to permitting the development of rigorous inventories of ecosystems.

A second implication refers to the necessity of there being a baseline when understanding processes of vegetation dynamics, whether they tend toward irreversible deterioration or toward processes of salvageable recovery. The precise description of the condition of the vegetation types and their tendencies are now fundamental to the ability to reverse undesirable processes or at least to document systematically those vegetation types vulnerable in the long and short terms.

It should be highlighted that, within the previous framework, the range of deterioration experienced by each vegetation type across its organizational levels varies drastically—generally known as the resilience mechanism. It is, for example, known that dry tropical deciduous forests are relatively more susceptible to certain deterioration processes than humid temperate perennial forests (Rzedowski 1978; Durán 2004). For this reason, it is fundamental to know each type of vegetation within an organized framework and hence to denote the resilience capacity of each. In Mexico, there is still a preponderance of dictating natural resource management and conservation policies regarding all vegetation type as functionally equal. As a result, an innovative and eventually fruitful policy such as sustainable forestry or payment for the provision of environmental water services is adequate for one type of vegetation but not for others as each has its own nature in terms of resilience capacity; all of which underscore the urgent need to develop uses and policies responsive to the particularities of each vegetation type. For such an aim, a single and rigorous system is crucial (Peterson 2002; Walker et al. 2002).

An additional advantage included in SECLAVEMEX is the possibility to bridge the gap between the various levels of biological organization and their eventual cartographical expression. This aspect is doubtless fundamental and far from exhausted. The analysis it provides is, moreover, sophisticated, given the complex nature of the vegetation patterns prevalent in Mexico.

Even so, it is from an organized outline that legends expressing vegetation distribution patterns are developed. Based on the various grouping levels, congruent work scales can be derived, permitting the analysis of said patterns whether for purely academic ends or for management, conservation, and restoration purposes. Among the academic aspects, there stands out the lack of any reference framework for, among others, biogeographical, palaeoecological, synecological, and ecological niche studies. In terms of applied aspects, there is an increasingly urgent need for a single inventory of Mexican vegetation types in their diverse organizational levels with a corresponding cartographical expression so as to corroborate unequivocally the baseline for territorial planning or, as it is better known in Mexico, the land use planning (—Ordenamiento Ecológico Territorial‖). Thematic mapping of some 60 distinct themes is currently necessary in order to substantiate the tasks of the —Ordenamientos Ecológicos Territoriales‖. No less than half of these themes in their diagnostic, characterization, integration, analytical, prognostic, or projection phases rely on vegetation-type maps to determine the condition, the tendency, and the potential uses for a territory, hence rendering crucial the availability of a SECLAVEMEX replete with a congruent territorial expression. Without such a system, inter-regionally incompatible OETs will continue to be developed and will be impossible to nest in macroregional outlines.

One of the applications worthy of deep reflection is the theme of conservation. To date, conservation schemes in Mexico have followed three lines: the establishment of various categories of Áreas Naturales Protegidas (ANP—protected natural areas), the declaration of Unidades de Manejo de Vida Silvestre (UMAS—wildlife management units) and the environmental services payment schemes (PSA —pago por servicios ambientales). These three approaches have been successful in some aspects, all the while being questionable and even contradictory in certain regions for various reasons: ANPs are, by their very nature, predominantly exclusive of cultural capital and tend toward a low level of operative efficiency in some highly populated regions. UMAS seem destined to address proposals which, in the main, never move beyond adding to an increasingly large number of demands for infinity on infinitely small spaces. Finally, the PSA programme exists within an erroneous conceptual framework, which, on the one hand, gives relative importance to a vegetation type from which goods or services are derived, ignoring that said product prevails through the functional connection of a group of vegetation types, and, which, on the other hand, incentivizes assigning a purely economic value to nature, largely ignoring the fact that what it aims to preserve has had its functional integrity excellently maintained locally by those who hold to principles beyond commercialization, namely cultural and historical values. In the light of previous framework, it is hoped that SECLAVEMEX will be able to contribute to the development of *ad hoc* conservation policies for all vegetation types, policies designed with the specific characteristics of each region of widely diverse countries such as Mexico in mind. Analyses of gaps or —ecological niches‖ need to be addressed in the short term such that policies strengthening existent conservation efforts can be derived—policies which can assist in correcting or reorienting conservation projects no longer deemed successful (Boege 2008).

New initiatives such as REED+ for the mitigation of climate change are also important given that the conceptualization and classification of as well as coherent nomenclature for —forests‖, for example, remain unresolved issues, not only in Mexico, but also in a large part of the world, thus providing another opportunity to test the relevance of SECLAVEMEX as much in the ecological definition of —forests‖ as in their territorial delimitation.

Mexico, with one of the world's most important emerging economies, with education centres ranking among the best in the work and with its extraordinary cultural and natural heritage, deserves a vegetation classification system rigorous in its conceptualization and simple in its application partly in order to maximize existent knowledge and partly to minimize costs. This goal was, without a doubt, an additional cost-benefit strength afforded by SECLAVEMEX. There are multiple benefits stemming from the use of an inclusive system, congruent with local reality, which draws on the recapitulation of findings from preceding works as way of reorienting for future experiences. In Europe, for example, thanks to the fact that the phytosociological approach to the study of vegetation was taken over a century ago there exists today a congruent strategy for prioritising management and conservation strategies for natural resources (http://www.rednatura2000.ue). A decade ago, Canada and the USA took the same path and, although they are still a long way from achieving what exists in Europe, they do have a very similar and compatible scheme (FGDC 2006). Mexico can start with SECLAVEMEX, improving it and developing it into a solid territorial expression of vegetation types, or it can remain on the verge of being able to deepen nationwide knowledge and thus of being able to generate information which would, in real time, contribute to natural- and cultural-resource management and conservation strategies and policies. This task is not the responsibility of a single academic discipline, institution, or social group. On the contrary, it is a collective effort based on agreement and consensus, concepts which are the axis of development in Mexico, a country crying out for new ways to reconcile growing social demands on all that an natural world, ample though under increasing pressure, has to offer.

This book represents the first in a series of eight additional currently underway, the aim of which is to compile a mesoscale inventory of Mexican vegetation types with corresponding cartographical representation. This goal could be achieved within fifteen years and would be possible with massive participation on the part of institutions, researchers, and students. The conceptual framework, however, was the fundamental challenge, and it is SECLAVEMEX, as presented in this contribution, which defines it.

Appendices

See Appendices A.1, A.2, A.3, A.4, A.5, A.6, A.7 and A.8.

Appendix A.1 Common levels and criteria of four classification systems in Mexico

Author(s)	Criteria used in their categories	Level (nomenclature)	Application of the criteria and levels
Miranda and Hernández X. (1963)	1. Floristic	Level I (Vegetation types)	The selection criteria are not applied uniformly in defining each vegetation type
	2. Growth form		The nomenclature is clear, but with a prevalent use of traditional terms
	3. Function		
	4. Height		
	5. Spines		
	6. Habitat		
	7. Succulence		
Rzedowski (1978)	1. Growth form	Level I: (Vegetation types)	The selection criteria are not applied uniformly in defining each vegetation type
	2. Climate	Level II: (Other vegetation types)	It is a hierarchical proposal with two levels, but the nomenclature is not standardized
	3. Function		
	4. Floristic		
	5. Spines		
	6. Xeromorphy		
	7. Habitat		
	8. Substratum		

(continued)

A. Velázquez et al., *Standardized Hierarchical Vegetation Classification*, Geobotany Studies, DOI 10.1007/978-3-319-41222-1

Appendix A.1 (continued)

Author(s)	Criteria used in their categories	Level (nomenclature)	Application of the criteria and levels
COTECOCA (1994)	1. Growth form	Level I: (Vegetation types)	The first criterion is applied consistently in defining the large groups (e.g. forests, jungles, shrubland, and grassland)
	2. Height		The following criteria are applied inconsistently, thus making the nomenclature confusing
	3. Function		
	4. Leaf morphology		
	5. Spines		
	6. Succulence		
	7. Density		
	8. Development		
	9. Substratum		
	10. Management		
González-Medrano (2003)	a. Climate zone	Level I: (Climate zone)	Hierarchy and standardized use of criteria exist
	b. Zonality or azonality	Level II: (Vegetation zonal azonal)	Includes a zonality criterion which brings into consideration which renders system confusing to the point of incomprehensibility
	1. Growth form	Level III: (Bioma)	The nomenclature also relies on different bases and criteria
	2. Function	Level IV: (Series of formation)	
	3. Height	Level V: (Formations)	
	4. Texture and consistency of leaves and stems	Level VI: (Associations)	
	5. Land cover	Level VII: (Consociations)	

Appendix A.2 Criteria of selection SECLAVEMEX and their description

Criteria	Dominant attributes	Qualifier's definition and description of the dominant elements or attributes
Physiognomy	Forest	Dominated by woody stem plants above 5 m tall
Growth form	Shrubland	Dominated by plants with one of more woody or succulent stems. Usually of less than 5 m in height or arborescent or arbofrutescent plants over 5 m high
	Grassland	Dominated by plants with no woody base
	Non-vascular	Plants with limited or no vascular system
Climate	Humid temperate	Climates with a mean annual temperature between 5° and 18 °C and precipitation from 1,000 to >4,000 mm
	Dry temperate	Climates with a mean annual temperature between 5° and 18 °C and precipitation from 0 to 1,000 mm
	Humid tropical	Climates with a mean annual temperature above 18 °C and precipitation from 1,000 to >4,000 mm
	Dry tropical	Climates with a mean annual temperature above 18 °C and precipitation from 0 to 1,000 mm
	Cold	Type E (T) C (cold and temperate. Climates with mean annual temperatures between −2° and 5 °C and precipitation from 0 to 2,000 mm
	Euriclimatic	Cold, temperate, or tropical. Optional qualifier Euriclimatic
Phenology of the foliage	Deciduous	Plant communities where 75 to 100 % of the top stratum individuals lose their leaves during the dry season
	Subdeciduous	Plant communities where 50–75 % of the top stratum individuals lose their leaves
	Evergreen	Plant communities where 75–100 % of the superior-canopy individuals remain green year round
	Subevergreen	Plant communities where 25–50 % of the top stratum individuals lose their foliage during the dry season
Presence of spines or vertical spikes	Spiny	Plant community dominated by individuals with spines
	Spineless	Plant community dominated by individuals without spines
	Subspineless	Plant communities with a density superior to 30 % of individuals with spines but inferior to 70 %
Leaf characteristics	Needle-leaved	Needle-shaped leaf
	Clustered	A clustering of flexible sprouts over a small area
	Angustifolia	Flat, narrow leaf
	Caespitose	Turf like (developed through creeping stems)
	Scales-leaved	Scale-shaped leaf
	Broadleaved	Broad, flat leaf
	Linearifolia	Narrow leaf with parallel edges
	Megaphyllous	Leaf larger than ± 150 cm^2
	Microphyllous	Leaf smaller than ± 2.5 cm^2
	Rosetophilous	Leaf cluster, generally helically, forming a rosette
	Others	Various unconsidered leaf types

(continued)

Appendix A.2 (continued)

Criteria	Dominant attributes	Qualifier's definition and description of the dominant elements or attributes
Succulence	Succulent	Communities with a predominance of plants with succulent or fleshy stems, leaves, and roots
	Non-succulent	Community of non-succulent dominant plants, neither in leaves stems
Floristic composition	Family	Family is a mid-level taxon within the system of taxonomic categories containing one or more genera (The taxon "Family" can be dominant is some communities, for example herbs in grassland and Pinaceae in some temperate forests)
	Genus	The taxon "genus" can be dominant in some plant communities, for *Quercus* or *Pinus* in holm or pine forests
	Species	"Species" is the basic taxon with the hierarchical taxonomic system. Numerous plant communities can be discerned by the dominance of 1–3 or more species
Substratum type	*Gypsophile*	Plant community adapted to gypsum-rich soils
	Halophytic	Communities adapted to high-salinity soils
	Aquatic and subaquatic	Plants adapted to water covering the soil
	Others	Substrata which and be relevant, but which do form part of this work

Appendix A.3 Previous studies permitting categories and criteria suggested by SECLAVEMEX to be inferred

Case study	Level I	Level II	Level III	Level IV	Level V	Level VI	Level VII
	Bioma	Major formation	Formation	Subformation	Series of associations	Association	Subassociation
Low deciduous jungle in Kalakmul western Campeche. (Slope centre on light soil hills and NW to 200 and 250 m.s.n.m. M.a. T. 24.6 °C and precipitation 1,076 mm, Canopy 15 m. (Martínez and Galindo-L et al. 2002)	Forest	Humid tropical	Deciduous	Broadleaved	*Pseudobombax–Bursera*	*Pseudobombax ellipticum*-and *Bursera simaruba*	*Lysiloma latisiliqua*
Low deciduous jungle at the UNAM Chamela biology station Jalisco. M.a. T 24.6°C and Precip. 788 mm (Durán 2004)	Forest	Dry tropical	Deciduous	Broadleaved	*Apoplanesia–Lonchocarpus*	*Apoplanesia paniculata, Lonchocarpus constrictus*	
Quercus forest in the La Michilía biosphere reserve SE of Durango, semi-dry temperate climate. M.A.T. 11.6 °C. Precip. 789 mm. (González-Elizondo et al 1993)	Forest	Dry temperate	Deciduous	Broadleaved	*Quercus*	*Quercus obtusata, Q.crassifolia, and Arctostaphylos pungens*	
Holm forest. Cerro Huitepec Reserve, San Cristóbal de las Casas, Chis. E-NE of the hill/slope. Elev. 2,450–2,260 m. M.a.T. 14-15 °C. Precip. 1,300mm (Ramírez-Marcial et al. 1998)	Forest	Humid temperate	Deciduous	Broadleaved	*Quercus*	*Quercus* spp *(95 %) Arbutus xalapensis and Alnus acuminata*	
Mountain mesophyll forest in Sierra San Carlos, Tamps. At 1,300–1,400 MSNM. Precip. 734.8 mm (Luna et al. 2001)	Forest	Dry tropical	Evergreen	Linear and broadleaved	*Abies–Carya*	*Abies guatemalensis, Carya ovata, Quercus* spp. *Carpinus caroliniana*	

(continued)

Appendix A.3 (continued)

Case study	Level I	Level II	Level III	Level IV	Level V	Level VI	Level VII
	Bioma	Major formation	Formation	Subformation	Series of associations	Association	Subassociation
Mountain mesophyll forest in Chinantla, Oax. At 1,600–1,800 MSNM. M.a. T. 14–17 °C. Precip. 5,797.8 mm (Luna et al. 2001)	Forest	Humid temperate	Evergreen	Broadleaved	*Engelhardtia*		*Engelhardtia mexicana*
Spiny shrubland. Tepejí del Río. Altitude 2,250 m. M.a. T. 15–16 °C. Precip. 605–790 mm (Romero and Rojas 1991)	Shrubland	Dry temperate	Evergreen	Spiny	*Opuntia–Acacia*	*Opuntia lindheimeri, Acacia schaffneri* and *Mimosa aculeaticarpa* var. *biuncifera*	
Xeric shrubland. Nizanda, Itsmo de Tehuantepec, Oax. Altitud 100–700 m. m.a. T. 15 °C. Precip. 1,000 mm (Pérez-García et al 2001)	Shrubland	Humid tropical	Evergreen	Succulent	*Agave–Hechtia*	*Agave ghiesbreghtii, Hechtia caudata, H. rosea*	
Grassland. Lake Zirahúen, Mich. M. a.T. 10 °C. Precip. ±1,149 mm (Pérez-Calix 1996)	Herbaceous	Humid temperate	Perennial	*Angustifolia* —clustered		*Muhlenbergia gigantea*	
Grassland Sierra de San Carlos. Tamps. M.a.T. 23 °C. Precip. 734 mm (Briones 1991)	Herbaceous	Dry tropical	Deciduous	*Angustifolia* —caespitose		*Agrostis semiverticillata, Bothriochloa saccharoides var. torreyana, Danthonia spicata,* etc.	

Appendix A.4 Standardized hierarchical classification system for vegetation in Mexico

Hierarchical level				
Level I	II	III	IV	V
Denomination				
Bioma	Major formation	Formation	Subformation	Series of associations
Definition criteria				
Physiognomy	Climate type	Foliage phenology (Stratum height)	Morphology, presence of Spines and succulence	Genus level dominance
Forest	Temperate, humid, or cold	Evergreen	*Linearifolia*	*Abies*
				Pseudotsuga and Picea
			Needle-leaved	*Pinus*
			Scale leaf	*Cupressus*
				Juniperus
			Needle-leaved, scale-*linearifolia*	*Pinus and Juniperus*
				Pinus and Abies
			Broadleaved	*Quercus*
			Needle-leaved, broadleaved	*Pinus* and *Quercus*
		Subevergreen or subdeciduous	Broadleaved	*Quercus*
		Deciduous	Broadleaved	*Alnus*
				Quercus
	Temperate, humid, or dry	Evergreen	Spineless, broadleaved	*Quercus, Pinus and Juniperus*
				Quercus, Juniperus and Bouteloua
	Temperate or tropical humid	Evergreen and subevergreen	Broadleaved	*Liquidambar, Ostrya and Carpinus*
				Alfaroa, Alnus and Carpinus
				Liquidambar, Quercus and Podocarpus
			Megaphyllous	*Sabal, Brahea, Orbignya and Scheelea*

(continued)

Appendix A.4 (continued)

Hierarchical level				
Level I	II	III	IV	V
				Sabal, Orbignya and Scheelea
				Sabal, Orbignya and Brahea
				Scheelea, Sabal and Pseudophoenix
	Dry temperate	Evergreen	Needle-leaved	*Pinus cembroides and P. pinceana*
			Scale leaf	*Juniperus flaccida and Juniperus* sp.
		Subdeciduous or deciduous	Spiny	*Acacia, Prosopis and Pithecellobium*
			Broadleaved	
			Sub-spineless	
	Humid tropical	Evergreen	Needle-leaved or scale leaf	*Pinus–Juniperus*
			Needle-leaved, broadleaved	*Pinus– Quercus*
			Broadleaved	*Terminalia, Swietenia* and *Brosimum*
				Terminalia, Swietenia and *Brosimum*
				Pachira, Chrysobalanus and *Calophyllum*
				Billia, Clusia, Engelhardtia and *Meliosma*
				Ryzophora, Avicennia, Laguncularia and *Conocarpus*
		Subevergreen	Broadleaved	*Achras, Bucida* and *Brosimum*
				Brosimum, Manilkara and *Bursera*
				Alseis, Pterocarpus, Carpodiptera and *Manilkara*

(continued)

Appendix A.4 (continued)

Hierarchical level				
Level I	II	III	IV	V
				Brosimum
				Haematoxylum, Metopium, Byrsonima and *Cresentia*
				Byrsonima, Curatella and *Cresentia*
		Subdeciduous	Broadleaved	*Enterolobium, Hymenaea* and *Orbignya*
				Hymenaea, Enterolobium and *Cedrela*
			Spiny	
	Dry temperate	Evergreen	Needle-leaved	*Pinus*
			Needle, broadleaved	*Pinus* and *Quercus*
			Broadleaved	*Quercus*
			Spiny	*Prosopis (Acacia)*
				Pithecellobium and Prosopis
		Subevergreen or subdeciduous	Broadleaved	*Brosimum, Ficus* and *Bursera*
			Spiny	*Cercidium, Pithecellobium, Olneya* and *Prosopis*
				Cercidium, Pithecellobium, Olneya and Prosopis
		Deciduous	Broadleaved	*Ceiba, Lysiloma* and *Cochlospermum*

(continued)

Appendix A.4 (continued)

Hierarchical level				
Level I	II	III	IV	V
				Plumeria, Bursera, Gyrocarpus and Spondias
				Piscidia, Lysiloma and *Cordia*
	Cold, temperate, or tropical (euriclimatic)	Evergreen	Broadleaved	*Taxodium, Platanus* and *Populus*
				Platanus, Populus, Salix and *Taxodium*
		Evergreen to deciduous	Not applicable	

Appendix A.5 Standardized hierarchical classification system for vegetation in Mexico

Hierarchical level				
Level I	II	III	IV	V
Denomination				
Bioma	Major formation	Formation	Subformation	Series of associations
Definition criteria				
Physiognomy	Climate type	Foliage phenology (stratum height)	Morphology, presence of spines and succulence	Genus level dominance
Shrubland	Humid temperate or cold	Evergreen to deciduous	Broadleaved, microphyllous	
	Dry temperate	Evergreen	Needle-leaved	*Pinus, Juniperus*
				Pinus
			Squamipholiate-leaf	*Juniperus*
			Broadleaved	*Quercus*
				Adenostoma, Arctostaphylos, Ceanothus, Quercus, etc.
				Quercus, Arctostaphylos, Cotoneaster, Adenostoma, etc.
			Microphyllous	*Larrea, Fouquieria, Cercidium, Ambrosia*, etc.
				Larrea, Flourencia, etc.
				Larrea
			Rosetophilous	*Yucca*
			Succulent, rosetophilous and/or spiny	*Agave, Hechtia, Dasylirion* and *Yucca*
				Agave, Fouquieria, Pachycormus, Machaerocereus, etc.
				Agave, Bergerocactus, Dudleya, etc.
				Agave, Hechtia, etc.
				Agave
				Agave, Hechtia, etc.
				Agave, Hechtia y/o Dasylirion, etc.

(continued)

Appendix A.5 (continued)

Hierarchical level				
Level I	II	III	IV	V
		Deciduous	Spineless	
			Subspineless	*Helietta, Neopringlea, Gochnatia*, etc.
				Helietta, Neopringlea, Cordia, Pithecellobium, etc.
				Helietta, Gochnatia, etc.
				Ipomoea, Bursera, Acacia, etc.
				Acacia, Eysenhardtia, Forestiera, Ipomoea,
			Spiny	*Condalia, Koeberlinia, Lycium, etc.*
				Acacia
				Cercidium, Acacia, Leucophyllum, Condalia, etc.
				Acacia
				Prosopis, Acacia, Mimosa, etc.
		Evergreen to deciduous	Succulent-subspineless	*Bursera, Jatropha, Pachycereus,* etc.
				Bursera, Jatropha, Ambrosia, etc.
			Succulent	*Fouquieria*
				Opuntia
				Cactáceas
				Cereus
				Myrtillocactus, Lemaireocereus, Escontria, etc.
				Pachycereus, Lophocereus, Cephalocereus, Opuntia, etc.

(continued)

Appendix A.5 (continued)

Hierarchical level				
Level I	II	III	IV	V
				Opuntia, Myrtillocactus, Carnegia, Stenocereus
				Cactáceas
				Lemaireocereus, Prosopis, Burseras, Karwinskia, etc.
				Fouquieria, Euphorbia, Lycium, Lophocereus, etc.
				Fouquieria, Yucca, Pachycereus and *Stenocereus*
				Fouquieria, Pachycereus, Opuntia, etc.
				Pachycereus, Machaerocereus, Lophocereus, Opuntia, etc.
			Microphyllous, broadleaved, and succulent	*Opuntia, Coccoloba, Bromelia*, etc.
				Ipomoea, Croton, Opuntia, etc.
				Prosopis, Larrea, Opuntia, Atriplex, etc.
				Hilaria, Opuntia, Ephedra, Eriogonum, etc.
	Cold, temperate, and tropical (euriclimatic)	Evergreen	Broadleaved to microphyllous	
		Deciduous	Broadleaved, microphyllous, succulent	

Appendix A.6 Vegetation study methods (Taken from Amador 2009)

Fixed surfaces or dimensions method
Minimum area. Quantitative approach for characterizing vegetation described by data. The sample size (collection of sampling units) must provide the most useful and accurate information possible. It is customary to define sampling units before carrying out any ecological study, given that these are regarded as the smallest area adequately representing the species composition within the community
In concordance with some authors, the following empirical values have been proposed:
Forest arboreal stratum: 200–500 m^2
Shrubland: 10–25 m^2
Forest herbaceous stratum: 50–200 m^2
Herbaceous: 30–100 m^2
Nested points method
The most usual method for defining the field area. A small area is recommended for initial consideration: 0.5 m^2 and that all species present should be noted. The area is doubled successively, and the additional species found in each duplication are noted
Simultaneously, a graph *i* constructed with the y- axis representing the number of species and the x-axis representing the samples area
The minimum area is the sampling area in which the curve turns almost horizontal
Square methods
Permits precise determination of species density, cover and frequency within the community and using this data to highlight the importance of each. This method does not necessarily imply square areas, and other forms can be used to delimit a constant and known area
Circular plane
Used widely in forest settings. The squared plane is used on flat terrain with abundant vegetation, whereas the rectangular plane is used where there are well-defined environmental gradients
Dimensional techniques (or variable dimensions)
There are various techniques using the distance between the plants or between the plants and a randomly chosen point. These techniques were developed and perfected for the study of the arboreal stratum of plant communities. They offer the advantage of not having delimited zones *a priori*. The two most commonly used techniques are those of points in a quadrant and random points
Points in a quadrant
Random points are located within the area to be sample. In many cases, however, it is appropriate to select points along a line or transect crossing the area to be described
A Cartesian coordinate axis is placed exactly at the sampling point dividing the zone of the site into four equal quadrants (hence the name of the technique). For each quadrant the distance from the centre of the coordinate axis to the nearest plant
Random points
Also require points selected at random. For each point, the plant closest to the centre of the quadrant is located. A line is traced from the point to the plant and then another line perpendicular to the first such that the second line divides the area into half. Then, the plant closest to the sampling point is located in the opposite half (that is the half not containing the first plant selected) with the distance between the two plants now being measured and recorded
Transect method
In some types of vegetation, many species present growth forms which make it difficult to distinguish the limits between individual plants. This characteristic renders inadequate the application of the surface (Fixed or Variable)
Canfield line
In 1941, R. Canield proposed the "line intercept method" to obtain cover estimates of species frequency in low-height communities. A measuring tape covering from 10 to 100 m distance is laid (the distance is based upon plant height). All plants intercepted by the tape are recorded in terms of species, and total cover along the tape. This method is repeated n times until all plant species of the community to be characterized are comprised

Appendix A.7 Relevé sheet with minimum information for a baseline study

Surveyor data and location of sampling unit						
Name				Institution		
Date		Municipality		State		
Latitude		Longitude		Elevation		
Map		Photograph/Image		Polygon Number		
Characterization data Landform						
	Mountain	Low Hills	Plain	Meseta	Valley	Other
Catenal	Crest	Slope	Piedmont	Valley		
Exposure		Slope	Concave slope		Convex slope	
Soil						
Type		Depth		Colour		
Texture		Structure		Drainage		
Substratum	Gypsophile	Halophytic	Hydrophilic	Other		
Level I Physiognomy Stratification						
	Height X	Coverage %	Classes of cover 1 = <1–1 % 2 = 2–5 % 3 = 6–25 % 4 = 26–50 % 5 = >50 %			
Arboreal						
Arbustive						
Herbaceous						
Flush						
Other						
Note Class coverage threshold values may vary according to the aim of the study						
Level II climate			Level III Phenology			
	Humid	Dry		Arboreal	Arbustive	Herbaceous
Tropical			Deciduous			
Temperate			Subdeciduous			
Cold	NA	NA	Evergreen			
Euriclimatic			Subevergreen			

(continued)

Appendix A.7 (continued)

Level IV Leaf Morphology

		Arboreal	Arbustive	Herbaceous	Other	
Presence of spines	Spiny					
	Spineless					
	Subspineless					
Leaf type						
	Needle-leaved					
	Clustered					
	Angustifoliate					
	Caespitose					
	Scale-leaf					
	Broadleaved					
	Linearifolia					
	Megaphyllous					
	Microphyllous					
	Rosetophilous					
	Others					
Succlence	Succulent					
	No-succulence					

Levels V-VII Floristic composition

	Collection number	Common name	Scientific name Family, genus, species	Stratum	Coverage	

Appendix A.8 Comparison among vegetation classification systems from Mexico

Hierarchical level						
I	II	III	IV			V
Bioma	Major formation	Formation	Subformation			Series of associations
Physiognomy	Climate-type	Phenology	Morphology presence of spines and succulence	Correspondence with previous proposals (original nomenclature in Spanish)	Author(s)	Level genus or family dominance
Forests	Humid or cold temperate	Evergreen	*Linearifolia*	Bosque de *Abies*	Rzedowski	*Abies*
				Bosque de abetos u oyameles	Miranda and HX	
				Bosque de oyamel (incluye ayarín & cedro)	I GEOGRAFÍA	
				Bosque de oyamel	INEGI	
				Bosque de ayarín	INEGI	*Pseudotsuga y/o Picea*
				Bosque de *Pseudotsuga* & de *Picea*	Rzedowski	
				Bosque templado alto de linearifolios	GLEZ-MED	
				Bosque templado mediano linearifolio ripario	GLEZ-MED	
				Bosque linearifolio	COTECOCA	
			Needle-leaved	Bosque de *Pinus*	Rzedowski	*Pinus*
				Pinares	Miranda and HX	

(continued)

Appendix A.8 (continued)

Hierarchical level						
I	II	III	IV			V
				Bosque de pino	I GEOGRAFÍA	
				Bosque de pino	INEGI	
				Bosque templado mediano aciculifolio gipsófilo	GLEZ-MED	
				Bosque templado mediano de aciculifolios	GLEZ-MED	
				Bosque aciculifolio	COTECOCA	
				Bosque claro aciculifolio sempervirente	UNESCO	
				Bosque de aciculifolios	INEGI	
			Scale leaf	Bosque de *Cupressus*	Rzedowski	*Cupressus*
				Cipresal	INEGI	
				Bosque de cedro	INEGI	
				Bosque de *Juniperus*	Rzedowski	*Juniperus*
				Bosque de enebros	Miranda and HX	
				Bosque de táscate	INEGI	
				Bosque de táscate	I GEOGRAFÍA	
				Bosque escuamifolio	COTECOCA	
				Bosque templado bajo de escuamifolios	GLEZ-MED	
			Needle-leaved, scale-linearifolia	Bosque aciculi-escuamifolio	COTECOCA	*Pinus & Juniperus*

(continued)

Appendix A.8 (continued)

Hierarchical level						
I	II	III	IV			V
				Bosque templado bajo de escuamiaciculifolios	GLEZ-MED	
				Bosque aciculi-linearifolio	COTECOCA	*Pinus & Abies*
				Bosque templado & subpolar sempervirente de coníferas	UNESCO	
				Bosque de coníferas	INEGI	
				Bosque de coníferas	Rzedowski	
			Broadleaved	Bosque de *Quercus*	Rzedowski	*Quercus*
				Bosque de encino	I GEOGRAFÍA	
				Bosque de encino	INEGI	
				Bosque esclerófilo	INEGI	
				Encinar	Miranda and HX	
				Bosque de roble de Veracruz	DINERSTEIN	
				Bosque de roble del centro de México	DINERSTEIN	
				Bosque mediano perennifolio de durifolios	GLEZ-MED	
				Bosque templado mediano de durifolios	GLEZ-MED	
				Bosque latifoliado sempervirente esclerófilo con lluvias de invierno	UNESCO	

(continued)

Appendix A.8 (continued)

Hierarchical level						
I	II	III	IV			V
				Bosque claro latifoliado sempervirente	UNESCO	
				Bosque esclerófilo perennifolio	COTECOCA	
			Needle-leaved, broadleaved	Bosque aciculi-esclerófilo	COTECOCA	*Pinus & Quercus*
				Bosque de pino & roble de la Sierra de Juárez	DINERSTEIN	
				Bosque de pino & roble de la Sierra Madre del Sur	DINERSTEIN	
				Bosque de pino & roble de la Sierra Madre Oriental	DINERSTEIN	
				Bosque de pino & roble de San Lucas	DINERSTEIN	
				Bosque de pino & roble transvolcánica de México	DINERSTEIN	
				Bosque de pino-encino	INEGI	
				Bosque de pino-encino (incluye encino-pino)	I GEOGRAFÍA	
				Bosque esclero-aciculifolio	COTECOCA	
				Bosque templado mediano de aciculidurifolio	GLEZ-MED	
				Bosque templado mediano de duriaciculifolios	GLEZ-MED	
				Bosque de encino-pino	INEGI	

(continued)

Appendix A.8 (continued)

Hierarchical level						
I	II	III	IV			V
				Bosque templado bajo de duriescuamifolios	GLEZ-MED	
				Bosques de coníferas & b. temp. de latifoliadas	DINERSTEIN	
		Subevergreen or subdeciduous	Broadleaved	Bosque de encino	INEGI	*Quercus*
				Bosque de encino	I GEOGRAFÍA	
				Bosque de *Quercus*	Rzedowski	
				Bosque esclerófilo	INEGI	
				Encinares	Miranda and HX	
				Bosque esclerófilo subcaducifolio	COTECOCA	
				Bosque mayormente sempervirente	UNESCO	
				Bosque templado latifoliado sempervirente estacional con lluvias en verano	UNESCO	
		Deciduous	Broadleaved	Bosque de *Alnus*	Rzedowski	*Alnus*
				Bosque templado mediano de durifolios	GLEZ-MED	*Quercus*
				Bosque templado mediano caducifolio	GLEZ-MED	

(continued)

Appendix A.8 (continued)

Hierarchical level						
I	II	III	IV			V
	Humid or dry temperate	Evergreen	Spineless, broadleaved	Bosque esclerófilo caducifolio	COTECOCA	
				Bosque bajo abierto	I GEOGRAFÍA	*Quercus, Pinus, Juniperus,*etc.
	Humid temperate or tropical	Evergreen or/and subevergreen	Broadleaved	Bosque bajo abierto	INEGI	*Quercus, Juniperus, Bouteloua*, etc.
				Bosque caducifolio	Miranda and HX	*Liquidambar, Ostrya, Carpinus*, etc.
			Megaphyllous	Bosque caducifolio	COTECOCA	*Alfaroa, Alnus, Carpinus*
				Bosque caducifolio	INEGI	
				Bosque mesófilo de montaña	I GEOGRAFÍA	
				Bosque mesófilo de montaña	INEGI	*Liquidambar, Quercus, Podocarpus*
				Bosque mesófilo de montaña	Rzedowski	
				Bosque bajo hidrófilo pinnado palmatifoliado	GLEZ-MED	*Sabal, Brahea, Orbignya, & Scheelea*
				Palmar	DINERSTEIN	
					I GEOGRAFÍA	
					COTECOCA	
					INEGI	*Sabal, Orbignya, Scheelea*

(continued)

Appendix A.8 (continued)

Hierarchical level						
I	II	III	IV			V
				Palmares	Rzedowski	*Sabal, Orbignya, Brahea*
					Miranda and HX	*Scheelea, Sabal, Pseudophoenix*
	Dry temperate	Evergreen	Needle-leaved	Bosque de *Pinus*	Rzedowski	*Pinus cembroides, P. pinceana*
			Scale leaf	Bosque de táscate	INEGI	*Juniperus flaccida, Juniperus*
		Subdeciduous or deciduous	Spiny	Bosque caducifolio espinoso	COTECOCA	*Acacia, Prosopis, Pithecellobium*
				Bosque denso espinoso	UNESCO	
				Bosque espinoso	INEGI	
				Bosque espinoso	Rzedowski	
			Broadleaved	Bosque claro deciduo por el frío, con árboles sempervirentes	UNESCO	
				Bosque claro extremadamente xeromórfico dominado por esclerófilas	UNESCO	
				Bosque deciduo por el frío con árboles (o arbustos) sempervirentes entremezclados	UNESCO	
				Bosque denso extremadamente xeromórfico dominado por esclerófilas	UNESCO	

(continued)

Appendix A.8 (continued)

Hierarchical level						
I	II	III	IV			V
	Humid tropical	Evergreen	Subspineless	Bosque claro deciduo por el frío, sin árboles sempervirentes	UNESCO	
				Bosque deciduo por el frío sin árboles sempervirentes	UNESCO	
				Bosque claro deciduo por la sequía	UNESCO	
			Subevergreen or subdeciduous	Bosque claro mayormente deciduo	UNESCO	
				Bosque mayormente deciduo	UNESCO	
				Bosque claro espinoso	UNESCO	
				Bosque sempervirente tropical & subtropical de coníferas estacional con lluvias en verano	UNESCO	*Pinus -Juniperus*
			Broadleaved	Bosque tropical & subtropical de coníferas	DINERSTEIN	
				Bosque tropical alto perennifolio	GLEZ-MED	*Terminalia, Swietenia, Brosimum, etc.*
				Bosque tropical bajo perennifolio	GLEZ-MED	
				Bosque tropical perennifolio	Rzedowski	
				Selva alta perennifolia	Miranda and HX	
				Selva alta perennifolia	INEGI	*Terminalia, Swietenia, Brosimum, etc.*

(continued)

Appendix A.8 (continued)

Hierarchical level						
I	II	III	IV			V
				Selva alta perennifolia	COTECOCA	
				Selva alta & mediana perennifolia	I GEOGRAFÍA	
				Selva baja perennifolia	I GEOGRAFÍA	*Pachira, Chrysobalanus, Calophyllum,* etc.
				Selva baja perennifolia	COTECOCA	
				Selva baja perennifolia	INEGI	
				Selva mediana a baja perennifolia	Miranda and HX	*Billia, Clusia, Engelhardtia, Meliosma,* etc.
				Selva mediana perennifolia	COTECOCA	
				Bosque tropical ombrófilo	UNESCO	
				Manglar	Rzedowski	*Ryzophora, Avicennia, Laguncularia & Conocarpus*
				Manglar	Miranda and HX	
				Bosque de manglares	UNESCO	
				Manglar	COTECOCA	
				Manglar	I GEOGRAFÍA	
				Manglares	DINERSTEIN	

(continued)

Appendix A.8 (continued)

Hierarchical level						
I	II	III	IV			V
				Manglar	INEGI	
		Subevergreen	Broadleaved	Bosque tropical alto subperennifolio	GLEZ-MED	*Achras, Bucida, Brosimum*, etc.
				Bosque tropical bajo subperennifolio	GLEZ-MED	
				Bosque tropical mediano subperennifolio	GLEZ-MED	
				Selva alta o mediana subperennifolia	Miranda and HX	
				Selva alta subperennifolia	COTECOCA	
				Selva alta subperennifolia	INEGI	*Brosimum, Manilkara, Bursera*,etc.
				Selva mediana subperennifolia	COTECOCA	
				Selva alta & mediana subperennifolia	I GEOGRAFÍA	*Alseis, Pterocarpus, Carpodiptera, Manilkara, etc.*
				Selva mediana subperennifolia	INEGI	*Brosimum*
				Selva baja subperennifolia	I GEOGRAFÍA	*Haematoxylum, Metopium, Byrsonima & Cresentia*
				Selva baja subperennifolia	COTECOCA	
				Selva baja subperennifolia	INEGI	
				Selva baja subperennifolia	Miranda and HX	*Byrsonima, Curatella & Cresentia*

(continued)

Appendix A.8 (continued)

Hierarchical level						
I	II	III	IV			V
				Bosque de *Byrsonima, Curatella & Cresentia*	Rzedowski	
		Subdeciduous	Broadleaved	Selva mediana caducifolia & subcaducifolia	I GEOGRAFÍA	*Enterolobium, Hymenaea, Orbignya*, etc.
				Bosque tropical & subtropical sempervirente estacional	UNESCO	
				Bosque húmedo de la Sierra Madre	DINERSTEIN	
				Bosque húmedo de Oaxaca	DINERSTEIN	
				Bosque húmedo de Tehuantepec	DINERSTEIN	
				Bosque húmedo de Yucatán	DINERSTEIN	
				Bosque húmedo tropical de hoja ancha	DINERSTEIN	
				Bosque montano de Veracruz	DINERSTEIN	
				Bosque tropical & subtropical semideciduo	UNESCO	
				Humedal de Quintana Roo	DINERSTEIN	
				Selva alta o mediana subcaducifolia	Miranda and HX	*Hymenaea, Enterolobium, Cedrela*, etc.
				Selva baja subcaducifolia	COTECOCA	
				Selva mediana subcaducifolia	COTECOCA	
			Spiny	Selva baja subperennifolia espinosa	COTECOCA	

(continued)

Appendix A.8 (continued)

Hierarchical level						
I	II	III	IV			V
	Dry tropical	Evergreen	Spiny	Selva baja subperennifolia subespinosa	COTECOCA	
				Bosque claro mayormente sempervirente	UNESCO	
		Subevergreen or subdeciduous	Broadleaved	Mezquital central de México	DINERSTEIN	
				Mezquital de San Lucas	DINERSTEIN	
				Mezquital occidental de México	DINERSTEIN	
				Mezquital	INEGI	*Prosopis* (*Acacia*)
				Mezquital (incluye huizachal)	I GEOGRAFÍA	
				Selva baja espinosa perennifolia	Miranda and HX	*Pithecellobium*, *Prosopis*, etc.
				Bosque tropical bajo subcaducifolio	GLEZ-MED	
				Bosque tropical mediano subcaducifolio	GLEZ-MED	
				Bosque tropical subcaducifolio	Rzedowski	*Brosimum*, *Ficus*, *Bursera*, etc.
				Selva baja subcaducifolia	INEGI	
				Selva mediana subcaducifolia	INEGI	
			Spiny	Selva baja subcaducifolia espinosa	COTECOCA	

(continued)

Appendix A.8 (continued)

Hierarchical level						
I	II	III	IV			V
				Selva mediana subcaducifolia espinosa	COTECOCA	
				Selva baja subcaducifolia subespinosa	COTECOCA	
				Selva baja espinosa	I GEOGRAFÍA	*Cercidium, Pithecellobium, Olneya, Prosopis,* etc.
				Selva baja espinosa	INEGI	
				Selva baja caducifolia espinosa	COTECOCA	
				Selva baja espinosa caducifolia	Miranda and HX	*Cercidium, Pithecellobium, Olneya, Prosopis,* etc.
				Selva mediana caducifolia espinosa	COTECOCA	
		Deciduous	Broadleaved	Selva mediana caducifolia	COTECOCA	
				Selva mediana caducifolia	INEGI	
				Selva alta caducifolia	COTECOCA	
				Bosque seco de Baja California	DINERSTEIN	
				Bosque seco de Balsas	DINERSTEIN	
				Bosque seco de Jalisco	DINERSTEIN	
				Bosque seco de Oaxaca	DINERSTEIN	

(continued)

Appendix A.8 (continued)

Hierarchical level						
I	II	III	IV			V
				Bosque seco de Sinaloa	DINERSTEIN	
				Bosque seco de Tamaulipas & Veracruz	DINERSTEIN	
				Bosque seco de Veracruz	DINERSTEIN	
				Bosque seco de Yucatán	DINERSTEIN	
				Bosque seco tropical de hoja ancha	DINERSTEIN	
				Bosques tropicales de latifoliadas	DINERSTEIN	
				Bosque deciduo por la sequía (tropical & subtropical)	UNESCO	
				Bosque tropical bajo caducifolio	GLEZ-MED	*Ceiba, Lysiloma, Cochlospermum,* etc.
				Bosque tropical caducifolio	Rzedowski	
				Bosque tropical mediano caducifolio	GLEZ-MED	*Plumeria, Bursera, Gyrocarpus, Spondias,* etc.
				Bosques tropical secos	GLEZ-MED	
				Selva baja caducifolia & subcaducifolia	I GEOGRAFÍA	
				Selva baja caducifolia	Miranda and HX	*Piscidia, Lysiloma, Cordia,* etc.

(continued)

Appendix A.8 (continued)

Hierarchical level						
I	II	III	IV			V
				Selva baja caducifolia	COTECOCA	
				Selva baja caducifolia	INEGI	
				Bosque subtropical ombrófilo	UNESCO	
	Cold, temperate or tropical (euclimatic)	Evergreen	Broadleaved	Bosque bajo a mediano perennifoliohalófilo	GLEZ-MED	*Taxodium, Platanus, Populus,* etc.
				Bosque de galería	INEGI	
				Otras comunidades leñosas	Rzedowski	
				Bosque bajo perennifolio hidrófilo	GLEZ-MED	
				Bosque mediano a alto perennifolioripario	GLEZ-MED	
				Selva de galería	INEGI	
				Vegetación de galería	I GEOGRAFÍA	*Platanus, Populus, Salix & Taxodium*
		Evergreen to deciduous	Not applicable	Bosque cultivado	INEGI	
				Plantación forestal	I GEOGRAFÍA	
Needle-leaved	Humid or cold temperate	Evergreen to deciduous	Broadleaved, microphyllous	Tundra de arbustos enanos, líquenes & musgos	UNESCO	
				Pantano turboso no elevado	UNESCO	
				Formación de pantano turboso de musgos con arbustos enanos	UNESCO	

(continued)

Appendix A.8 (continued)

Hierarchical level						
I	II	III	IV			V
				Pantano turboso elevado	UNESCO	
				Matorral deciduo por el frío	UNESCO	
				Matorral denso enano deciduos por el frío	UNESCO	
				Matorral claro enano deciduos por el frío	UNESCO	
	Dry temperate	Evergreen	Needle-leaved	Matorral de coníferas	I GEOGRAFÍA	*Pinus, Juniperus*
				Matorral de coníferas	INEGI	*Pinus*
				Matorral de *Pinus*	Rzedowski	
				Matorral claro sempervirente aciculifolio micrófilo	UNESCO	
				Matorral denso sempervirente aciculifolio micrófilo	UNESCO	
			Scale leaf	Matorral de *Juniperus*	Rzedowski	*Juniperus*
			Broadleaved	Matorral mediano esclerófilo	COTECOCA	*Quercus*
				Matorral mediano esclerófilo perennifolio	GLEZ-MED	
				Matorral alto esclerófilo	COTECOCA	
				Matorral bajo esclerófilo	COTECOCA	
				Matorral arborescente	COTECOCA	
				Matorral arbofrutescente	COTECOCA	
				Chaparral	I GEOGRAFÍA	*Adenostoma, Arctostaphylos, Ceanothus, Quercus*, etc.

(continued)

Appendix A.8 (continued)

Hierarchical level						
I	II	III	IV			V
				Chaparral	INEGI	*Quercus, Arctostaphylos, Cotoneaster, Adenostoma*, etc.
				Matorral enano mayormente sempervirente	UNESCO	
				Matorral denso enano sempervirente	UNESCO	
				Matorral claro enano sempervirente	UNESCO	
				Matorral de arbustos enanos sempervirente herbáceo	UNESCO	
				Matorral mediterráneo	DINERSTEIN	
			Microphyllous	Matorral desértico micrófilo	I GEOGRAFÍA	*Larrea, Fouquieria, Cercidium, Ambrosia*, etc.
				Matorral inerme	INEGI	*Larrea, Flourencia*, etc.
				Matorral inerme parvifolio	Miranda and HX	
				Matorral alto inerme	GLEZ-MED	
				Matorral mediano parvifolio	COTECOCA	*Larrea*
				Matorral mediano parvifolio crasicaulescente	COTECOCA	

(continued)

Appendix A.8 (continued)

Hierarchical level						
I	II	III	IV			V
			Rosetophilous	Matorral desértico micrófilo	INEGI	*Yucca*
				Izotal	COTECOCA	
			Succulent, rosetophilous and/or spiny	Izotales	Miranda and HX	
				Izotal	INEGI	
				Matorral desértico rosetófilo	I GEOGRAFÍA	*Agave, Hechtia, Dasylirion*y *Yucca*
				Matorral rosetófilo costero	I GEOGRAFÍA	*Agave, Fouquieria, Pachycormus, Machaerocereus*, etc.
				Matorral rosetófilo costero	INEGI	*Agave, Bergerocactus, Dudleya*, etc.
				Matorral con rosetófilos acaules	INEGI	*Agave, Hechtia*, etc.
				Matorral desértico rosetófilo	INEGI	*Agave*
				Crasi-rosulifolios espinosos	Miranda and HX	*Agave, Hechtia*, etc.
				Matorral crasirosulifolio espinoso	COTECOCA	
				Crasi rosulifolios	INEGI	*Agave, Hechtia y/o Dasylirion*, etc.

(continued)

Appendix A.8 (continued)

Hierarchical level						
I	II	III	IV			V
				Matorral alto rosetófilo espinoso	GLEZ-MED	
				Matorral bajo rosetófilo espinoso	GLEZ-MED	
		Deciduous	Spineless	Matorral mayormente deciduo	UNESCO	
				Matorral claro subdesértico deciduo	UNESCO	
				Matorral denso mayormente deciduo	UNESCO	
				Matorral denso enano facultativamente deciduos por la sequía	UNESCO	
				Matorral claro enano facultativamente deciduo por la sequía	UNESCO	
				Matorral denso enano obligatoriamente deciduos por la sequía	UNESCO	
				Matorral claro enano obligatoriamente deciduos por la sequía	UNESCO	
				Matorral claro subdesértico deciduo	UNESCO	
				Matorral denso enano & comunidades relacionadas	UNESCO	

(continued)

Appendix A.8 (continued)

Hierarchical level						
I	II	III	IV			V
			Subspineless	Matorral submontano	I GEOGRAFÍA	*Helietta, Neopringlea, Gochnatia*, etc.
				Matorral submontano	INEGI	*Helietta, Neopringlea, Cordia, Pithecellobium*, etc.
				Matorral alto subinerme	COTECOCA	*Helietta, Gochnatia*, etc.
				Matorral alto subinerme	GLEZ-MED	
				Matorral mediano subinerme	COTECOCA	
				Matorral mediano subinerme	INEGI	
				Matorral mediano subespinoso	Miranda and HX	
				Matorral bajo subespinoso	COTECOCA	
				Matorral deciduo por la sequía, con plantas leñosas sempervirentes entremezcladas	UNESCO	
				Matorral deciduo por la sequía, sin plantas leñosas sempervirentes entremezcladas	UNESCO	
				Matorral inerme o subinerme parvifolio	Miranda and HX	

(continued)

Appendix A.8 (continued)

Hierarchical level						
I	II	III	IV			V
				Matorral subtropical	I GEOGRAFÍA	*Ipomoea, Bursera, Acacia,* etc.
				Matorral subtropical	INEGI	*Acacia, Eysenhardtia, Forestiera, Ipomoea*
			Spiny	Matorral espinoso con espinas terminales	Miranda and HX	*Condalia, Koeberlinia, Lycium, etc.*
				Matorral espinoso con espinas laterales	Miranda and HX	*Acacia*
				Matorral espinoso tamaulipeco	I GEOGRAFÍA	*Cercidium, Acacia, Leucophyllum, Condalia,* etc.
				Matorral alto espinoso	GLEZ-MED	*Acacia*
				Matorral alto espinoso	COTECOCA	
				Matorral mediano espinoso	COTECOCA	
				Matorral bajo espinoso	COTECOCA	
				Huizachal	INEGI	
				Matorral espinoso	INEGI	*Prosopis, Acacia, Mimosa,* etc.
				Matorral espinoso tamaulipeco	INEGI	
		Evergreen to deciduous	Succulent-spineless	Matorral sarcocaule	I GEOGRAFÍA	*Bursera, Jatropha, Pachycereus,* etc.
				Matorral sarcocaulescente	COTECOCA	
				Matorral sarcocaule	INEGI	*Bursera, Jatropha, Ambrosia,* etc.
				Matorral oliocilindrocaule afilo	COTECOCA	

(continued)

Appendix A.8 (continued)

Hierarchical level						
I	II	III	IV			V
			Succulent	Cirio	INEGI	*Fouquieria*
				Chollal	INEGI	*Opuntia*
				Nopaleras	Miranda and HX	
				Nopalera	INEGI	
				Matorral de cactos de Guerrero	DINERSTEIN	Cactáceas
				Matorral de cactos del centro de México	DINERSTEIN	
				Matorral de cactos del norte de Sonora	DINERSTEIN	*Cereus*
				Cardonales, tetecheras, etc.	Miranda and HX	*Myrtillocactus, Lemaireocereus, Escontria,* etc.
				Cardonal	INEGI	
				Matorral crasicaule	I GEOGRAFÍA	*Pachycereus, Lophocereus, Cephalocereus, Opuntia*, etc.
				Matorral crasicaule	INEGI	*Opuntia, Myrtillocactus, Carnegia, Stenocereus*
				Matorral crasicaule	COTECOCA	Cactáceas
				Matorral crasicaulescente	COTECOCA	

(continued)

Appendix A.8 (continued)

Hierarchical level						
I	II	III	IV			V
				Matorral arbocrasicaulescente	COTECOCA	*Lemaireocereus, Prosopis, Burseras, Karwinskia*, etc.
				Matorral sarcocrasicaule de neblina	I GEOGRAFÍA	*Fouquieria, Euphorbia, Lycium, Lophocereus*, etc.
				Matorral sarcocrasicaule	I GEOGRAFÍA	*Fouquieria, Yucca, Pachycereus & Stenocereus*
				Matorral sarco-crasicaule	INEGI	*Fouquieria, Pachycereus, Opuntia*, etc.
				Matorral sarcocrasicaulescent e subinerme	COTECOCA	
				Matorral sarco-crasicaule de neblina	INEGI	*Pachycereus, Machaerocereus, Lophocereus, Opuntia*, etc.
				Matorral alto crasicaule espinoso	GLEZ-MED	
				Matorral mediano crasicaulescente espinoso	COTECOCA	
				Matorral bajo crasifolio	COTECOCA	

(continued)

Appendix A.8 (continued)

Hierarchical level						
I	II	III	IV			V
			Microphyllous—broadleaved, succulent	Vegetación de dunas	COTECOCA	
				Vegetación de dunas costeras	INEGI	*Opuntia, Coccoloba, Bromelia,* etc.
				Vegetación de dunas costeras	Miranda and HX	*Ipomoea, Croton, Opuntia, etc.*
				Vegetación de desiertos arenosos	INEGI	*Prosopis, Larrea, Opuntia, Atriplex,* etc.
				Vegetación de desiertos arenosos	I GEOGRAFÍA	*Hilaria, Opuntia, Ephedra, Eriogonum,* etc.
	Cold, temperate or tropical (euclimatic)	Evergreen	Broadleaved, microphyllous	Matorrales medianos hidrófilos	GLEZ-MED	
				Matorral claro latifoliado sempervirente	UNESCO	
				Matorral denso latifoliado sempervirente	UNESCO	
				Matorral mayormente sempervirente	UNESCO	
				Matorral claro subdesértico mayormente sempervirente	UNESCO	

(continued)

Appendix A.8 (continued)

Hierarchical level						
I	II	III	IV			V
				Matorral claro subdesértico mayormente sempervirente	UNESCO	
		Deciduous	Broadleaved, microphyllous, succulent	Matorrales medianos halófilos	GLEZ-MED	
				Matorrales medianos gipsófilos	GLEZ-MED	
				Matorrales bajos a mediano arenícolas	GLEZ-MED	
				Matorrales medianos hidrohalófilos	GLEZ-MED	
Herbaceous	Cold or temperate	Perennial	Angustifolia—cluster	Pastizal amacollado	GLEZ-MED	
				Pastizal amacollado abierto	COTECOCA	
				Pastizal amacollado de montaña	GLEZ-MED	
				Pastizales de montaña	GLEZ-MED	
				Pastizal amacollado de montaña con arbustos esparcidos	GLEZ-MED	
				Pradera de alta montaña	I GEOGRAFÍA	*Festuca, Calamagrostis, Trisetum, Arenaria*, etc.
				Pradera de alta montaña	INEGI	
				Vegetación graminoide alta	UNESCO	
				Vegetación graminoide alta con una sinusia leñosa que cubre más del 10 %	UNESCO	

(continued)

Appendix A.8 (continued)

Hierarchical level						
I	II	III	IV			V
				Vegetación graminoide alta con una sinusia leñosa que cubre más del 10 %	UNESCO	
				Vegetación graminoide alta con una sinusia leñosa que cubre más del 10 %	UNESCO	
				Zacatonal	DINERSTEIN	
				Zacatonales	Miranda and HX	*Stipa*, *Muhlenbergia* & *Festuca*, etc.
			Angustifolia—broad-leaved	Páramo	DINERSTEIN	
				Tundra alpina	COTECOCA	
				Tundra alpina de México	DINERSTEIN	
				Vegetación de páramos de altura	Miranda and HX	*Arenaria, Draba, Muhlenbergia*, etc.
	Dry temperate	Perennial	*Angustifolia*—cluster	Pastizal amacollado arbofrutescente	COTECOCA	Gramineae
				Pastizal amacollado arborescente	COTECOCA	
				Vegetación graminoide alta con sinusia arbórea que cubre el 10–40 %	UNESCO	
				Vegetación graminoide alta con una sinusia arbórea que cubre del 10–40 %	UNESCO	

(continued)

Appendix A.8 (continued)

Hierarchical level						
I	II	III	IV			V
			Angustifolia—Caespitose	Vegetación graminoide alta con una sinusia arbórea que cubre del 10–40 %	UNESCO	
				Herbazal templado	GLEZ-MED	
				Pastizal cespitoso	GLEZ-MED	
				Pastizal cespitoso, con arbustos	GLEZ-MED	
				Pastizal del centro de México	DINERSTEIN	
				Pastizal del oriente de México	DINERSTEIN	
				Pastizal mediano abierto	COTECOCA	
				Pastizal mediano arbofrutescente	COTECOCA	
				Pastizal mediano arborescente	COTECOCA	
				Pastizal natural	INEGI	*Bouteloua*, *Buchloë*
				Pastizal natural (incluye pastizal-hizachal)	I GEOGRAFÍA	*Bouteloua*, *Trichachne*, *Hilaria*, *Aristida*, etc.
				Pastizales	Miranda and HX	*Bouteloua*, *Muhlenbergia*, *Lycurus*, etc.
				Vegetación graminoide baja	UNESCO	
				Vegetación graminoide baja sin sinusia leñosa	UNESCO	

(continued)

Appendix A.8 (continued)

Hierarchical level						
I	II	III	IV			V
				Vegetación graminoide de altura intermedia	UNESCO	
				Vegetación graminoide de altura intermedia prácticamente sin sinusia leñosa	UNESCO	
				Vegetación graminoide mesofítica intermedia a baja	UNESCO	
		Deciduous	Angustifolia—microphyllous	Pastizal natural-Huizachal	INEGI	*Bouteloua*, *Sporobolus*, *Muhlenbergia*, *Acacia*,etc.
				Restinga (dunas costeras)	DINERSTEIN	
				Vegetación de dunas costeras	I GEOGRAFÍA	*Ipomoea*, *Sporobolus*, *Uniola*, *Chamaecrista*, etc.
				Vegetación de desiertos áridos arenosos	Miranda and HX	*Panicum*, *Sporobolus*, *Prosopis*, etc.
	Humid or dry tropical	Perennial	Angustifolia—broadleaved	Sabana	INEGI	
				Sabana	Miranda and HX	*Andropogon*, *Paspalum*, *Trichachne*, etc.
				Sabana	I GEOGRAFÍA	*Andropogon*, *Paspalum*, *Setaria*, etc.

(continued)

Appendix A.8 (continued)

Hierarchical level						
I	II	III	IV			V
				Sabana	COTECOCA	
				Sabana de Tabasco & Veracruz	DINERSTEIN	
				Sabana de Tehuantepec	DINERSTEIN	
				Vegetación graminoide alta con una sinusia arbustos (sabana arbustiva)	UNESCO	
			Angustifolia—Megaphyllous	Sabana de palmas de Jalisco	DINERSTEIN	Gramineae
				Sabana de palmas de Veracruz	DINERSTEIN	
				Vegetación graminoide de altura intermedia con una sinusia abierta de plantas empenachadas	UNESCO	
	Cold, temperate or tropical (euclimatic)	Perennial	Angustifolia	Herbazal angustifolio inundado	GLEZ-MED	
				Herbazal inundable angustifolio	GLEZ-MED	
				Humedal del centro de México	DINERSTEIN	
				Pastizales inundables con o sin árboles esparcidos	GLEZ-MED	
				Popal	INEGI	
				Popal	Rzedowski	*Thalia, Calathea,* etc.

(continued)

Appendix A.8 (continued)

Hierarchical level						
I	II	III	IV			V
				Popal	Miranda and HX	*Calathea, Thalia, Leersia*, etc.
				Popal-tular	I GEOGRAFÍA	*Calathea, Thalia, Heliconia & Typha*
				Tular	INEGI	
				Tular & carrizal	Rzedowski	*Typha, Phragmites, Scirpus*, etc.
				Tulares, carrizales	Miranda and HX	*Typha, Phragmites, Scirpus*, etc.
			Broadleaved	Comunidad de agua dulce arraigada	UNESCO	
				Comunidad de agua dulce que flota libremente	UNESCO	
				Herbazal flotante	GLEZ-MED	
				Herbazal inundable latifoliado	GLEZ-MED	
				Herbazal latifoliado inundado	GLEZ-MED	
				Herbazal ripario	GLEZ-MED	
				Herbazal ripario	GLEZ-MED	
				Herbazales hidrófilos	GLEZ-MED	
				Otras comunidades herbáceas anfibias & subacuáticas	Rzedowski	*Acrostichum, Ammannia, Crinum*, etc.
				Pastizal hidrófilo con arbustos escuamifolios	GLEZ-MED	

(continued)

Appendix A.8 (continued)

Hierarchical level						
I	II	III	IV			V
				Vegetación flotante	Rzedowski	*Lemna*, *Spirodela*, *Wolffia*, etc.
				Vegetación hidromórfica de agua dulce	UNESCO	
				Vegetación hidrófila dulceacuícola	GLEZ-MED	
				Vegetación de galería	INEGI	
				Vegetación sumergida	Rzedowski	*Cabomba*, *Ceratophyllum*, *Myriophyllum*, etc.
		Deciduous	Angustifolia, broadleaved and succulent	Agrupaciones de halófitos	Miranda and HX	*Suaeda*, *Atriplex*, *Distichlis*, etc.
				Pastizal amacollado arenícola	GLEZ-MED	
				Pastizal amacollado halófilo	GLEZ-MED	
				Pastizal cespitoso gipsófilo	GLEZ-MED	
				Pastizal cespitoso halófilo	GLEZ-MED	
				Pastizal edáfico	GLEZ-MED	
				Pastizal gipsófilo	INEGI	
				Pastizal halófilo	INEGI	
				Pastizal halófito abierto	COTECOCA	
				Pastizal halófito arbofrutescente	COTECOCA	
				Pastizales halófilos amacollados	GLEZ-MED	

(continued)

Appendix A.8 (continued)

Hierarchical level						
I	II	III	IV			V
				Pastizales halófilos cespitosos	GLEZ-MED	
				Pastizales templados edáficos	GLEZ-MED	
				Vegetación halófita	Rzedowski	*Suaeda*
				Vegatación halófita	COTECOCA	
				Vegetación halófila	INEGI	*Atripex*, *Suaeda*, *Batis*, *Frankenia*, etc.
				Vegetación halófila & gipsófila	I GEOGRAFÍA	*Suaeda*, *Atriplex*, *Allenrolfea*, *Ambrosia*, etc.
			Angustifolia	Pastizal cultivado	INEGI	Gramineae
				Pastizal cultivado	I GEOGRAFÍA	Gramineae
				Pastizal inducido	INEGI	
				Pastizal inducido	COTECOCA	Briofitas
				Pastizal inducido	I GEOGRAFÍA	
Non-vascular	Humid cold or tempera te	Perennial	Not applicable	Formación de pantano turboso de musgos con arbustos enanos	UNESCO	
				Tundra mayormente de briofitas	UNESCO	Líquenes
				Tundra mayormente de líquenes	UNESCO	
				Área sin vegetación aparente	I GEOGRAFÍA	

SECLAVEMEX provides sound criteria to organize, name, and describe vegetation types consistently with previous contributions, hierarchically and rigorous in its application. Original vernacular names are given in Spanish to avoid misunderstandings when translating them

References

Amador GAL, Torres R, Aguirre L, Ortega JM, Villaseñor G, Ayala G (2001) Manifestación de impacto ambiental L.T. Agustín Millán II-Volcán Gordo (415 kV), Edo. de México. CFE-UMICH.

Amador GA (2009). Estructura de la vegetación y unidades de respuesta hidrológica. Tesis doctoral. Facultadde Ciencias Naturales. Universidad Autónoma de Querétaro.134 pp.

Anónimo (2003) Perfect scientific classification is first possible when one know? Document of Microsoft Word. http://www.fs.fed.us/emc/nfma/includes/revisedviabilityalternative.pdf.

Boege E (2008) Patrimonio Biocultural de los pueblos indígenas de México. Hacia la conservación in situ de la biodiversidad y agrodiversidad en los territorios indígenas, Instituto Nacional de Antropología e Historia, Comisión Nacional para el Desarrollo de los Pueblos Indígenas, México.

Braun-Blanquet J (1979) Fitosociología. Bases para el estudio de las comunidades vegetales. Blume, Madrid.

Briones O (1991) Sobre la flora, vegetación and fitogeografía de la Sierra de San Carlos, Tamaulipas. Acta Botánica Mexicana. Instituto de Ecología A.C. Pátzcuaro, Michoacán México, 16: 15–43.

Canfield R (1941) Application of the line interception method in sampling range vegetation. Journal of Forestry 39: 388–394.

Carpenter JR (1940) The grassland biome. Monogr. 10: 617–687.

COTECOCA (1994) Revegetación and reforestación de las áreas ganaderas en las zonas áridas and semiáridas de México. Comisión Técnico Consultiva para la definición de los Coeficientes de Agostadero. México.

Don_Faber-Langendoen (2004) HRWG: Compilation of Issues and Decisions. Microsoft Words document.

Drude O (1890) über Prinzipen in der Unterscheidung von Vegtations formationen, erläutert an der Cetraleuropäischen. Flora/Engler Bot. Jahrb., 11.

Durán ME (2004) Estructura, diversidad and mortalidad del componente arbóreo en un mosaico ambiental de Chamela, México. Tesis de Doctorado. Universidad Nacional Autónoma de México. México, D. F.

Ehrlich R, Ehrlich A (2005) One with Nineveh: Politics, Consumption, and the Human Future. Island Press, Washington, D.C.

FGDC (Federal Geographic Data Committee) (2006) National Vegetation Classification Standard, Version 2 – Working Draft. Vegetation Subcommittee. Federal Geographic Data Committee, August 2006.

González-Elizondo S, González-Elizondo M, A Cortés O (1993) Vegetación de la Reserva de la Biosfera - La Michilía, Durango, México. Acta Botánica Méxicana. Instituto de Ecología, A.C. Pátzcuaro, Michoacán. México. 22: 1–104.

A. Velázquez et al., *Standardized Hierarchical Vegetation Classification*, Geobotany Studies, DOI 10.1007/978-3-319-41222-1

González-Medrano F (2003). Las comunidades vegetales de México.Propuesta para la unificación de la clasificación y nomenclatura de la vegetación de México. Instituto Nacional de Ecología, México, DF.

González-Elizondo S, González-Elizondo M. Márquez M (2007) Vegetación y ecorregiones de Durango. Plaza y Valdez Editores. México, D. F.

González CO, Giménez de Azcárate J, García-Pérez J, Aguirre-Rivera R (2007) Flórula vascular de la Sierra de Catorce and Territorios adycentes, San Luis Potosí, México. Acta Botánica Mexicana. Enero, número 078. Instituto de Ecología, A.C. Pátzcuaro, México: 1–38.

Herrera A (1869) Apuntes para la geografía botánica de México. Naturaleza 1: 81–86.

Humboldt A von (1807) Essay sur la Géographie des Plantes. Par Al. de Humboldt and Bonpland, redigé par Al. de Humboldt. Levrault, Schoell et Cie, Paris.

INEGI (Insitituto Nacional de Estadística Geografía e Informática). (2005) Estados Unidos Mexicanos. Principales Avances en Materia Cartografica. VIII Conferencia Cartografica Regional de las Naciones Unidas para America, New York.

Jardel P, Graf S, Gómez G (2002). Manejo forestal comunitario y conservación en la Reserva de la Biosfera Sierra de Manantlán. V Congreso Nacional de Áreas Naturales Protegidas de México. Asociación Mexicana para la Conservación de las Áreas Naturales Protegidas A.C. y Gobierno del Estado de Jalisco. Guadalajara, Jalisco, México.

Koeppen W (1948) Climatología. Fondo de Cultura Económica. México, D.F.

Luna I, Velázquez A, Velázquez E (2001) México. In: Kappelle, M. and A. D. Brown. Bosques 427 nublados del neotrópico. Instituto Nacional de Biodiversidad. Fundación Agroforestal de 428 Argentina. Comité Holandés de la Unión Mundial para la Conservación de la Naturaleza.

Linnæus C (1735–1770) Systema naturae per regna tria naturae, secundum classes, ordines, genera, species, cum characteribus, differentiis, synonymis, locis. con 13 ediciones corregidas and aumentadas. https://www.google.com/search?tbm=isch&q=Linn%C3%A6us+C+1735 &hl=es&authuser=0#imgrc=zyt2-t4YmsH8CM%3A.

Lott EJ (1993) Annotated checklist of the vascular flora of the Chamela Bay Region, Jalisco, Mexico. Occasional Papers of the California Academy of Sciences 148: 1–60.

Martens M, Galeotti H (1842) Mémoire sur les fougères du Mexique et considerations sur la géographie de cette countrée. Mém. Acad. Sci. Bruxelles 15: 1–99.

Martínez E, Galindo-Leal C (2002) La vegetación de Calakmul, Campeche, México: Clasificación, Descripción and Distribución. Bol. Soc. Bot. México 71: 7–32.

Mayr E (2000) Biology in the twenty-first century. BioScience, 50(10): 895–897.

Miranda F, Hernández XE (1963) Los tipos de vegetación de México y su clasificación. Bol. Soc. Bot. Méx. 28: 29–179.

Ochotorena I (1918) Las regiones geográfico-botánicas de México. Bol. Soc. Mex. Geog. Est. V (8): 221–231.

Pedrotti F (2013) Plant and vegetation mapping. Springer, Heidelberg.

Pérez-Calix E (1996) Flora and vegetación de la cuenca del Lago de Zirahuén, Michoacán, México. Flora del Bajío and de Regiones Adyacentes. Fascículo Complementario XIII. Instituto de Ecología. Pátzcuaro, Michoacán. México.

Pérez-García EA, Meave J, Gallardo C (2001) Vegetación and flora de la región de Nizanda, Istmo de Tehuantepec, Oaxaca, México. Acta Botánica Mexicana. Instituto de Ecología A.C. Pátzcuaro, Michoacán. México, 56: 19–88.

Peterson D (2002) Contagious disturbance, ecological memory, and the emergence of landscape pattern. Ecosystems, 5:329-338.

Ramírez-Marcial N, Ochoa-Gaona S, González-Espinosa M, Quintana-Ascencio PF (1998) Análisis florístico and sucesional en la estación biológica Cerro Huitepec, Chiapas, México. Acta botánica Méxicana. Instituto de Ecología A.C. Pátzcuaro, Michoacán. México, 44: 59–85.

Rangel SR & Zenteno ECR. (1991) Estudio florístico de la Región de Huehuetoca, Estado de México. Acta Botánica Mexicana. Instituto de Ecología A.C. Pátzcuaro, Michoacán. México, 14: 33–58.

Riley D, Young A (1966) World Vegetation. Cambridge University Press. London.

Rivas-Martínez S, Sánchez-Mata D, Costa M (1999) North American Boreal and Western Temperate Forest Vegetation. (Syntaxonomical Synopsis of the potential natural plant communities of North America, II) Itinera Geobot. 12: 5–316.

Romero S, Rojas E (1991). Estudio florístico de la región de Huehuetoca, Estado de México. Acta Bot. Mex. 14: 33-57.

Rzedowski J (1978) Vegetación de México. Limusa. México, D.F.

Strasburger E, Noll F, Schenck H, Schimper AFW (1982) Tratado de Botánica. Omega, 7ª ed., Barcelona.

Velázquez A, Medina C, Reygadas D (2010a) Síntesis geobotánica de las sierras Ajusco-Chichinautzin. Investigación ambiental. Ciencia and política pública, Instituto Nacional de Ecología-SEMARNAT. 2(1) http://www.revista.ine.gob.mx/article/view/90.

Velázquez A, Picón AR (2010b) Los retos para lograr un bienestar ambiental duradero. Interciencia 35 (2) 5.

Walker B, Carpenter S, Anderies J, Abel N, Cumming G, Janssen M, Lebel L, Norberg J, Peterson G, Pritchard R (2002) Resilience management in social-ecological systems: a working hypothesis for a participatory approach. Conservation Ecology, 6(1): 14.

MIX
Papier aus verantwortungsvollen Quellen
Paper from responsible sources
FSC® C105338

If you have any concerns about our products,
you can contact us on
ProductSafety@springernature.com

In case Publisher is established outside the EU,
the EU authorized representative is:
Springer Nature Customer Service Center GmbH
Europaplatz 3, 69115 Heidelberg, Germany

Printed by Libri Plureos GmbH
in Hamburg, Germany